Cambridge Monographs in African Archaeology
72
Series Editors: John Alexander, Laurence Smith and Timothy Insoll

Memory and the Mountain

Environmental Relations of the Wachagga of Kilimanjaro and Implications for Landscape Archaeology

Timothy A. R. Clack

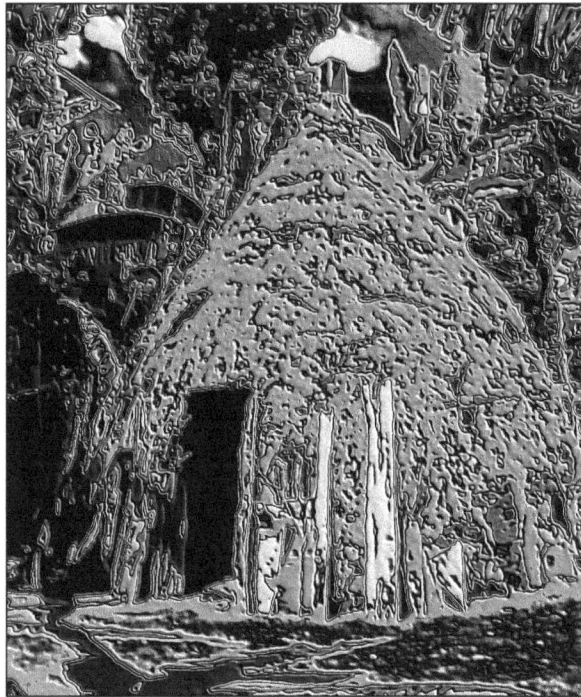

I0086058

BAR International Series 1679
2007

Published in 2016 by
BAR Publishing, Oxford

BAR International Series 1679

Cambridge Monographs in African Archaeology 72
Memory and the Mountain

ISBN 978 1 4073 0117 4

© T A R Clack and the Publisher 2007

COVER IMAGE *Representation of a traditionally constructed Wachagga hut distorted to recognise, in some fashion, the influence of memory on experience and identity.*

Front Digital Montage: Digitally enhanced selection of photographs and artistic representations of Uchagga from the colonial and contemporary period which attempts to convey some of the memories embodied in the memoryscape.

Rear Digital Montage: Distorted depiction of the front digital montage which attempts to convey the corruption, signification and resonation of memories, identities and experiences.

The author's moral rights under the 1988 UK Copyright,
Designs and Patents Act are hereby expressly asserted.

All rights reserved. No part of this work may be copied, reproduced, stored,
sold, distributed, scanned, saved in any form of digital format or transmitted
in any form digitally, without the written permission of the Publisher.

BAR Publishing is the trading name of British Archaeological Reports (Oxford) Ltd.
British Archaeological Reports was first incorporated in 1974 to publish the BAR
Series, International and British. In 1992 Hadrian Books Ltd became part of the BAR
group. This volume was originally published by Archaeopress in conjunction with
British Archaeological Reports (Oxford) Ltd / Hadrian Books Ltd, the Series principal
publisher, in 2007. This present volume is published by BAR Publishing, 2016.

Printed in England

BAR
PUBLISHING

BAR titles are available from:

BAR Publishing
122 Banbury Rd, Oxford, OX2 7BP, UK
EMAIL   info@barpublishing.com
PHONE   +44 (0)1865 310431
FAX   +44 (0)1865 316916
www.barpublishing.com

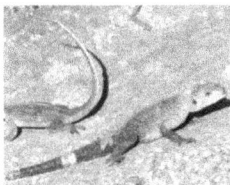

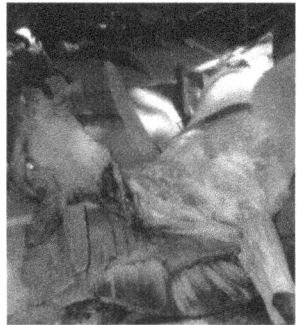

To My Closest Ancestors

CONTENTS

Acknowledgements   ix

Chapter 1: Introduction   1

Chapter 2: Memoryscapes: Collapsing Memory into Landscape   4
    Approaches to Archaeological and Anthropological Landscapes   4
    Cultural Memoryscapes   5
    Religious Memoryscapes   8
    Memoryscapes and the Classical Theories of Religion   11

Chapter 3: Being-Somewhere: Mythic Memoryscapes of the Wachagga   15
    Setting the Scene: Wachagga of Kilimanjaro   15
    Exploring African Landscapes: Some Issues   15

Chapter 4: Choreographies of Religious Experience   18
    Syncretism: Definitional Concerns   18
    Missionary Activity on Kilimanjaro   19
    Contemporary Religious Identities and Meanings: Evidence of Integrative Missionisation   21
    Contemporary and Historical Funerary Rites   29

Chapter 5: Physicality of Kibo   34
    Reverence and Irreducibility   34
    Traditions, Ritual Memory and Cosmology   34
    Protection, Security and Shelter   37

Chapter 6: Mountain Bounty: The Landscape as Provider   40
    Provisions   40
    Water and Irrigation Furrows   44
    Geophagic Consumption of Soil   46
    Tourist Pilgrimages   47
    Local Environmentalism   48

Chapter 7: Respecting the Past: Elders and Ancestors   50
    Origins of People and the World   50
    Elders, Patriclan Relations and Learning   51
    Ancestral Spirits and Supernatural Agency   52
    Human Remains and Shrines to the Ancestors   54
    Rituality: Slaughter and the Landscape   57

Chapter 8: Inscribing the Land: Language, Place and Ownership   62
    Place and Language   62
    Rights of Ownership: Dwellings and Land Tenure   63

Chapter 9: Specific Features of Some Memoryscapes in Uchagga   65
    Sieny Catchment Forest   65
    Kisumbe Reservation   71
    Kifinuka   75

Chapter 10: Concluding Remarks: Memoryscapes and Archaeology   78

Appendices   81
References   95

LIST OF FIGURES

Figure 1   16
Map showing the Kilimanjaro Region in relation to the countries of East Africa

Figure 2   24
Photograph of urovirovi (tangle-weed) on concrete for contrast

Figure 3   25
Photograph of Kibo from position in the Maharo study area

Figure 4   26
Table showing the orientation of local churches in study areas

Figure 5.1   27
Photograph of Nyikamisa Lya Mrima Pentecostal Church, Machame

Figure 5.2   27
Photograph of Muoamini Lutheran Church, Machame

Figure 5.3   28
Photograph of Mkuu Roman Catholic Church, Maharo

Figure 5.4   28
Photograph of Old Mkuu Roman Catholic Church, Maharo

Figure 6   32
Photograph of sacrificial portions of meat placed upon freshly cut banana leaves

Figure 7   33
Photograph of funerary kisusio preparation

Figure 8   38
Photograph of Laban caves

Figure 9   39
Photograph of Nkowoyo Kyalia bolt hole

Figure 10.1   41
Photograph of a traditional msonge constructed with banana leaves

Figure 10.2   42
Photograph of a traditional msonge constructed with grasses

Figure 11   43
Photograph of a modern dwelling sited in the kihamba

Figure 12   47
Symbolic dimensions and associations of geophagic and reproductive processes

Figure 13   49
Photograph of environmental cooking technologies from the Machame area

Figure 14.1   55
Photograph of male mbuoni from Marangu (note human humerus at base of shrine)

Figure 14.2   56
Photograph of female mbuoni from Marangu

Figure 15   57
Photograph of libation pouring instrument discarded at mbuoni

Figure 16   60
Photograph of Shao patriclan ifumvu la mkuu

Figure 17   61
Map depicting the location of the Shao patriclan ifumvu la mkuu and the various
political/administrative units incorporated in the 'dialectic-soil zones'

Figure 18   66
Sketch map of places of power within the Sieny Catchment Forest area

Figure 19.1   67
Photograph of Daraja la Mungu from position of high relief

Figure 19.2   68
Photograph of Daraja la Mungu from on the feature

Figure 20.1   68
Photograph of Nkyeku cave

Figure 20.2   69
Photograph of Nkyeku cave illustrating the markings of significance

Figure 21   70
Photograph of Kukeleweta Chemka

Figure 22   70
Photograph of Lake Nyumba ya Mungu

Figure 23   72
Photograph of the Kisumbe Reservation

Figure 24.1   73
Photograph looking-out from the top of Kinukamori Waterfall

Figure 24.2   74
Photograph looking-back from the top of Kinukamori Waterfall

Figure 25   76
Photograph from position in Kifinuka

Figure 26   82
Map showing locations of the three interview transects on the slopes of Mount Kilimanjaro

ACKNOWLEDGEMENTS

This work would never have been completed without the support and insight of my doctoral supervisors Prof. Timothy Insoll and Prof. Julian Thomas. Moreover for their assistance and thought-provoking dialogue special appreciation is also owed to other academics at the University of Manchester including Dr Marcus Brittain, Dr Chantal Conneller, Dr Sarah Croucher, Dr Melanie Giles, Dr Sian Jones, Dr Stephanie Koerner, Mr Chris Perkins and Prof. Raymond Tallis. I would like to note my appreciation of the help offered from Dr Paul Lane, Dr Rachel McLean and Dr Stephanie Wynn-Jones of the British Institute in Eastern Africa, and Prof. Chris Gosden, Dr Renee Hirschon and Prof. Peter Mitchell of the University of Oxford. Similarly Dr Bertram Mapunda, Mr Daryl Stump, Mr Mattias Tagseth and Dr Ludger Wimmelbücker are thanked for their magnanimous assistance. In addition the fellows and students of St Peter's College, University of Oxford and St Anselm Hall with Canterbury Court, University of Manchester are thanked for graciously providing the motivation, time and facilities to pursue the venture. Funding for research was received from the British Institute in Eastern Africa, Marcia Pointon Bursary Fund, Royal Anthropological Institute, Sir John Zochonis Fund and the Sir Richard Stapeley Educational Trust. COSTECH are acknowledged for permitting the research. In keeping with the requests of COSTECH and various regional offices copies of permissions and letters of introduction have been included in the appendices. My partner and family deserve special thanks for their support and understanding. Most importantly of all, however, I would like to note my appreciation of the Wachagga of Kilimanjaro who permitted me into their lives with unfaltering hospitality.

Then they began to climb and they were going to the East it seemed, and then it darkened and they were in a storm, the rain so thick it seemed like flying through a waterfall, and then they were out and Compie turned his head and grinned and pointed and there, ahead, all he could see, as wide as all the world, great, high, and unbelievably white in the sun, was the top of Kilimanjaro. And he knew that there was where he was going.

Ernest Hemingway *The Snows of Kilimanjaro*.

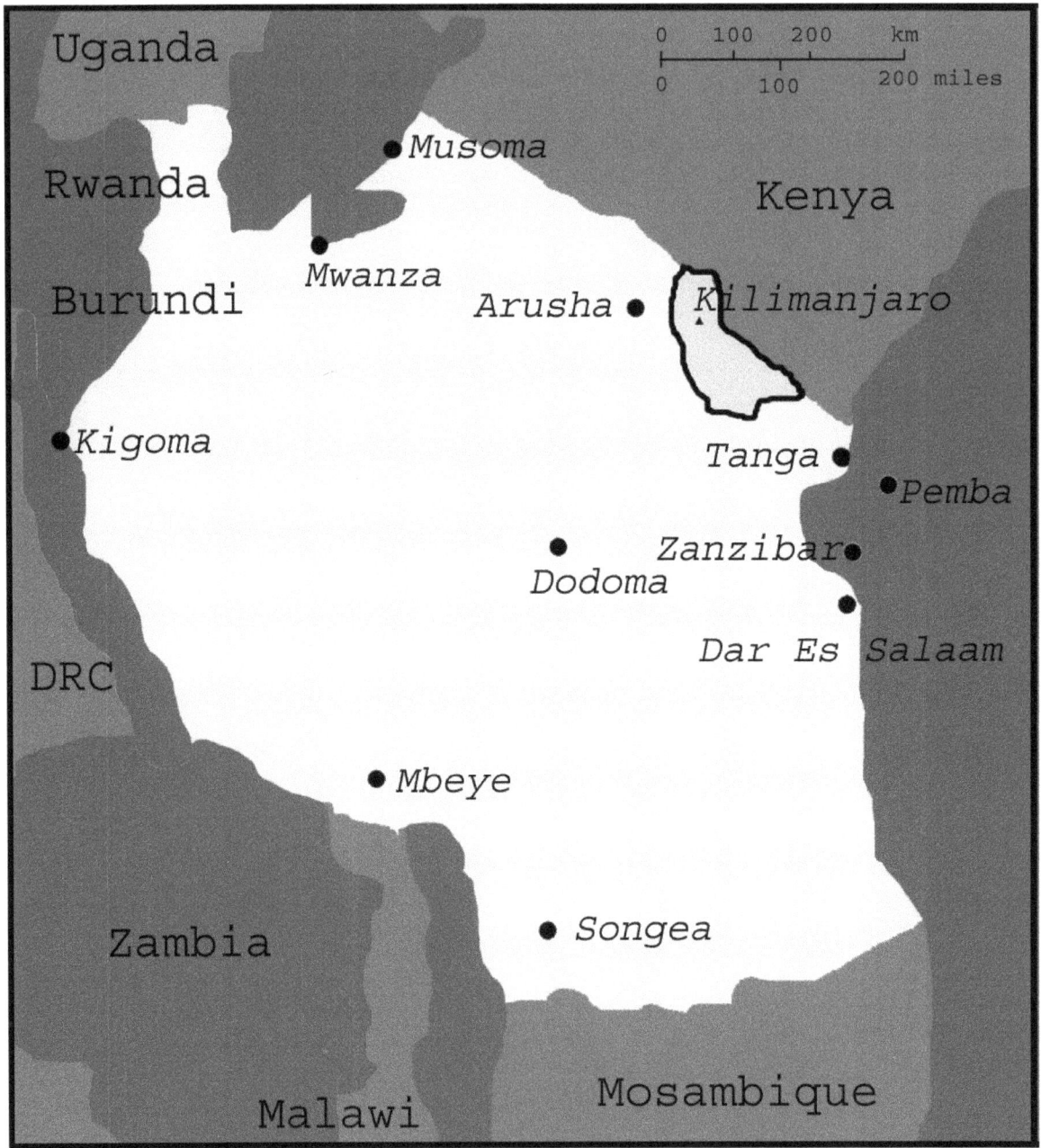

CHAPTER ONE

INTRODUCTION

"I am part of all that I have met" wrote Tennyson in *Ulysses*. The intriguing condition of humanity and its composition through relations, memories, histories, meanings and symbols could not have been captured with more beauty or accuracy. In this sense the person is made rather than born. This book considers the relationships between memory, experience and landscape from insights gained conducting ethnographic research. In so doing this investigation into the memoryscape might be labelled an 'archaeological ethnography' for not only is it an ethnography produced by an archaeologist, it was conducted with archaeological applications in mind. The uses of ethnography in archaeology are far too numerous to mention but in the main such ethnography has been produced by ethnographers and has not been prepared with any consideration of archaeological purposes.

It should be stated from the outset that one is not questioning the quality or sensitivity of typical ethnographies – they fulfil the research objectives of the ethnographers that produce them. The issue is that if archaeologists are using ethnography to enhance the 'archaeological imagination' through inferential and analogical methods then they require a greater understanding of ethnographic techniques and in an ideal world the opportunities and motivations to conduct their own ethnographic research. It has long been realised that archaeology and anthropology are related (if not actually the same) disciplines but cross-fertilisation of concepts, methods and insights are too uncommon. Moreover the dialogue that does exist seems to be rather one-sided with archaeology engaging anthropology. Archaeology should be seen as vital to the anthropological enterprise as it offers time depth and appreciations of how the past is recapitulated in the present. Moreover archaeology has to theorise very different worldviews which must, one suspects, fall within the anthropological remit. The dialogue between archaeologists and anthropologists should therefore be facilitated by both parties. One means by which anthropologists might be encouraged to engage archaeological discourses is with the production of archaeological ethnographies. This is an area of mutual concern and both parties should be motivated to explore the uses and applications of such research. Rather than pursuing 'ethno-archaeology' perhaps some form of 'archaeo-ethnography' would be better suited to certain archaeological applications.

The motivation of this ethnographical research was the theorisation of some issues involved in landscape archaeology. There has been considerable attention given to the religious dimensions of archaeological landscapes.

The purpose of this study is to augment such appreciations by theorising the complex environmental relations of a contemporary culture. Such theorisation enables recognition of the wealth of human experience that is archaeologically unknowable. This is not to contend that archaeologists have been completely ignorant of such insights but rather that without a deliberate comparative exercise the scale of the omission might be underplayed. This study serves primarily as a measure of the interpretable but it also highlights some complexities and concerns that may influence enhanced comprehension of human-landscape relations. Elements of these relations can be usefully conceptualised as the memoryscape.

What is a memoryscape? The memoryscape characterises elements of certain religious experience. Essentially it is the intersection of memory, emotion and being-somewhere and is a feature of the experiential world that may evoke a response related to the numinous in certain contexts. The memoryscape could be characterised as a hardwired yet subjective phenomenon that is of species-wide potential but is culturally particular in expression. All humans dwell within a memoryscape. It should be explicitly recognised from the outset that the theorisation concerning the memoryscape can be understood in two fashions. In the cognitive paradigm it is (albeit sophisticated) representationalist for any memoryscape must ultimately be 'represented' in the mind. In the phenomenological paradigm the memoryscape is part of the context or Background of experience. The subtleties and archaeological applications of these various paradigms is well beyond the scope of this venture (see Gosden 1994; Renfrew 1998; Thomas 1996 for further discussion) but it is worth noting their significance in recent landscape archaeologies.

With high degrees of precision memoryscapes are optimally presented in their contemporary forms. Myths and rituals coordinate and shape experience and should be conceptualised and theorised as doing so. For a myth "simultaneously imposes an order, accounts for the origin and nature of that order, and shapes people's depositions to experience that order in the world around them" (Bell 1997: 21, in Insoll 2004a: 123). Studies using myths should contextualise the mythology by involving ethnographies and knowledge of the surroundings (2004a: 127). Thus myth and ritual might also provide the means to "recapitulate the past in the present" (Ricoeur 1985: 17, in Insoll 2004a: 129) although temporal and cultural change must be acknowledged. Hence one can present the mythic memoryscape but only speculate about its origin and development in the past. To conceptualise and represent a memoryscape one must gather the memories of informants. The ethnographic informer can provide access to their memories, and such methods and endeavours have been recognised as a fertile arena for cultivating knowledge (see Teski and Climo 1995; Climo and Cattell 2002a). Indeed *memory repertoires* are a homogenising societal force instrumental in the inheritance and endowment of memories and meaning. Thus ethnography can question which memories are shared, how these memories were learned/transferred, the selectivity of memory and so on (Climo and Cattell 2002b: ix). It will be demonstrated that certain memoryscapes are religious in character. Obviously this is not to suggest that religion equates with memory or landscape but rather that in certain contexts there does seem to be some fashion of relationship.

Some have suggested that the task of defining religion is "doomed from the outset" (Flood 1999: 43). Perhaps such a position is too pessimistic the task of definition/comprehension though somewhat overwhelming should still be engaged with. To attain some level of consistency within archaeological discourses the definition to be utilised herein will be that tentatively provided elsewhere "a system of collective, public actions which conform to rules ('ritual') and usually express 'beliefs' in the sense of a mixture of ideas and predispositions" (Insoll 2001b: 2) but it should also be added that religions invoke the experience of the sacred. Although one could criticise the all-encompassing and arbitrary nature of this definition, it does serve in its inclusiveness. Nonetheless, understanding religion requires the recognition that it will include irrational, intangible, and indefinable elements (Insoll 2004a: 7; 2004c: 1). Ritual is the focusing lens of the sacred (see Smith 1980) that has been described as involving a homogenisation of emotion, experience, knowledge, movement and communication (Insoll 2004a: 10). All humans seem given to think religiously (e.g. about the world and ones place in) it albeit with different cultural attunements and understandings. Such consideration begs the question: does everyone have the capability of religious experiences, even if they reject the possibility of religious experience? It could be contended that all humans are religious beings including those that would describe themselves as atheist. The historic and contemporary evidence suggests this to be the case. This includes, for example, 'non-religious' subjects who behave in religious ways or experience the world through religious cultures. It is interestingly to note the word religion is actually derived from the Latin 'religio' meaning 'to unite or bind together' (Wilson 2002: 220). It will be demonstrated that this original meaning is particularly apt when one considers how *memoryscapes* – the cosmological framework born out of the correspondence between landscape and memory – unite and embed individuals and collectivities with/in the environment.

Various social scientists have quite rightly stressed that terms such as religion, spiritual, supernatural and sacred have taken on a smug condescension that relates back to the subjects of study i.e. subjects who believe in superstitions demonstrated to be false are somehow backward (e.g. Lohmann 2003: 118). Moreover these terms carry multiple meanings and dense cultural baggage. Supernatural, for instance, is not a particularly profitable term in that it sets up natural as the binary opposite. Thus such terminology propagates and reinforces power relations concerning control that were founded in the colonial 'civilising' enterprise. Interestingly the other subordinate dualistic pair of natural is unnatural. Thus supernatural becomes associated with the meanings of unnatural – deviancy and unfitness.

Some have suggested the anthropologists of religion should think in terms of an 'enhanced natural' rather than supernatural (Sered 2003: 218). This deliberation begs the question: are these terms able to function with sufficient relativism to avoid the pitfalls of ethnocentrism and exoticism? Lohmann (2003: 119) notes that the standard ethnographic practice to minimise such ethnocentric misrepresentation is to consider the cosmological premises of other cultures as indigenously 'true'. He suggests that to improve such a method one should consider the sacred or supernatural as existing within the minds of ethnographic subjects. Implicit within his argument is the recognition that such cosmological premises are ostensibly false to the ethnographer (2003: 120). It is also worthy of note that the Western cultural conceptualisation of the supernatural might not be based on the science/anti-science dichotomy forwarded by Lohmann. The high degree of overlap and fusion between the natural and the supernatural especially in societies that give primacy to the scientific method have been illustrated (Jindra 2003). Numerous contemporary pseudo-religious examples from the scientific West can be cited including Marxism, spiritual movements such as Scientology, Human Potential, Christian Science, Transcendental Meditation, *Star Trek* fandom, Unitarian Universalism, and 'reflexive spirituality'. Reflexive spirituality, for instance, is the incorporation of scientific rationality with the search for transcendent meaning (see Besecke 2001; Roof 1999). Illustrious mainstream scientists that have delved into theological arenas include Albert Einstein and Stephen Hawkings. Thus as Jindra (2003: 165) concludes "[s]cience and supernaturalism are often intertwined" and "the construction of meaning is basic to humanity".

This book is essentially divided into two sections. The following chapter documents some of the commonest archaeological and anthropological approaches to landscapes and subsequently explores the theory of the memoryscape. The proceeding chapters present ethnographic findings relating to the memoryscapes, and in particular the supernatural dimensions of environmental relations, of the Wachagga of Kilimanjaro, Tanzania. Various themes are explored including the choreography of religious experience, Christianisation, missionisation and indigenisation of place, and the temporality, meaningfulness and physicality of the landscape. These themes embody the types of knowledge about the past which, it can be confidently assumed, are beyond the interpretive boundaries of most archaeologies.

CHAPTER TWO

MEMORYSCAPES: COLLAPSING MEMORY INTO LANDSCAPE

APPROACHES TO ARCHAEOLOGICAL AND ANTHROPOLOGICAL LANDSCAPES

Anthropological and archaeological discourses have frequently concentrated on the theorising of landscape (e.g. Bender 1993; Rockman and Steele 2003; Rossignol and Wandsnider 1992; Ucko and Layton 1999). The default position seems to be a systemic understanding that landscape equates with background. Landscape is thus "a framing convention which informs the way the anthropologist brings the study into 'view'" and "refers to the meaning imputed by local people to their cultural and physical surroundings" (Hirsch 1995: 1). Thus the ethnographic frame has been noted to culturally exclude some subjects. In response some anthropologists have practised the manufacture of "textual pictures" (Lund 1998: 125). In so doing landscapes have been directly engaged with as the experiential ground of culture. They can be objects, subjects, representations, and experiences and the inherent meaning within each conception frequently merge and overlap (Lemaire 1997: 5; Thomas 2001: 166). To refer to any landscape as *terra incognito* is thus a misnomer. In fact usually it is wiser to speak of *landscapes* rather than the singular landscape as there are usually multiple associative value systems applicable to an area (e.g. Bender 1992). Landscape tradition is a manifestation of culture. Landscape provides contextual knowledge that informs the lived experience (Mintz 1989: 791). One copes with the environment through history and memory.

Landscape is produced through the interplay of the experiencing individual and their perspective. The Western landscape tradition sees the environment 'spectated' or 'viewed' with other spectators being distributed throughout the scene. Landscape is usually conceptualised as the stimuli/aesthetic exhibit (Weiner 1994: 2). This is problematic especially if it is realised that others within the scene also experience the landscape from different perspectives (Lund 1998: 125). Landscape is perceived. This perception is made intelligible through spontaneous pre-understandings. It should be remembered that perception is by no means restricted to the visual field. Loomis *et al* (1992) detail how the visually-impaired successfully wayfind by memorising and recalling routes by following auditory/tactile/olfactory rather than visual landmarks. Hence the experience of landscape is informed by other sensory fields such as smell, sound, texture, taste and atmosphere (Darvill 1999: 107). Furthermore the aesthetic conception of landscape has additional drawbacks. The aesthetic is predicated upon value-judgements that run contrary to the anthropological

practice of contrast annotation (Gow 1994: 22). The Western understandings of landscape particularise it through visual practice. Nevertheless landscape is made according to the local circumstances. The scenario of the non-indigene in Amazonia where the 'view' is occluded by vegetation exemplifies this (see Gow 1995). Landscape is made intelligible through indigenous practice. The concept of "anthropologyland" (Dirks 1994: 483) optimally describes the landscape of local practice understood through fieldwork (Hirsch 1995: 2). The world is frequently recognised as exhibition but this is by no means a universal phenomenon. Although things are exhibited through their physical visibility it must be acknowledged that they are "made by people as they live their relations to other people and to things" (Harvey 1996: 42). Landscape is not only the signifier within a symbolic/sign system but it is also the referent and integral to the message (Morphy 1995: 186).

> "[H]uman beings adapt not to their environment but to their ideas about it, even if effective adaptation requires a reasonably close correspondence between reality and how it is perceived" (Trigger 1989: 261)

Phenomenological accounts have noted that through dwelling in the landscape it becomes part of our constitution "just as we are part of it" (Ingold 1993: 158). Dwelling is constituted through tasks. These tasks are habitually conducted rather than consciously confronted (1993: 162). The notion of the taskscape has been introduced to better conceptualise the circumstances of dwelling. According to Ingold (1993: 162) the landscape consists of forms and features. This is different to the taskscape which is optimally understood as the circumstances of dwelling. It is important to note that landscape and taskscape do not reside in opposition. Indeed "landscape is the congealed form of taskscape" (1993: 162). The taskscape is the activity-biased sphere of the landscape the features of which are produced and exists alongside it. One is a participant in the perception of the temporality of the taskscape and landscape rather than a spectator (1993: 159).

The dwelling perspective correctly proposes that landscape is a produced subject and is therefore not an object. Human existence involves Being-somewhere (Thomas 1996: 83). This involves the production of place through the inhabitation of spaces by cultural bodies. Space is rendered meaningful through involvement in experience. Landscape relates to experiential space through networks of *de-severance* and *directionality*. De-severance is the process whereby things are recognised as distinct entities in order to facilitate their relational placement. Directionality is characterised by one's involvement and attachment with things e.g. emotional closeness (1996: 83-4). Thomas highlights that certain places are bodily understood and thus recede from explicit concern and through inhabitation they are ascribed an innate form of closeness (1996: 86). Moreover relationships with places are like relationships with other human beings (Jagger 1985: 221). Ethnographies inform that perceptions of, and values attached to, landscape "encode values and fix memories to places that become sites of historical identity" (Stewart and Strathern 2003: 1). Identities thus concern the conflation of two notions – *memory* and *place*. Stewart and Strathern (2003: 2) note that place and memory are crucial transducers that bring the community into mutual alignment. Indeed landscape provides a context for the negotiation of place, memory and community (2003: 3).

CULTURAL MEMORYSCAPES

Recent archaeological and anthropological literature concerning landscape argue for what might be termed 'cultural memoryscapes' suggesting that landscape can be optimally comprehended as a fusion of culture and nature. In such discourse warnings have been forwarded that stress the current academic tendency to over socialise space (e.g. McGlade 1999: 461). The physical landscape has features that must be acknowledged. Of course these will be subjectively made intelligible but nevertheless they will restrict, offer different perspectives, have ecological consequences and so on. Some studies thus dislocate social practice from the ecological and natural milieu which is misrepresentative (Benton 1994: 45). McGlade suggests a compromise study of human ecodynamics that better relates subjects to their environmental setting.

> "What we are arguing for is caution against any representation of cultural schemata (social, symbolic, structural) that distances itself from the temporalities of lived experience – not those simply related to bodily experience, but particularly the insertion of such experiences in natural and physical phenomena. Without such embedding, we are in danger of constructing fictive landscapes – landscapes whose only points of reference are residual networks of meaning structures" (McGlade 1999: 461)

The separation of the cultural and the natural is a modernist trait that erroneously sets up humanity as the arbitrator of reality. Heidegger refers to this post-Enlightenment conceptualisation as the 'age of the world picture' (Heidegger 1977b: 119-29). The notion of place better rectifies such misunderstanding. Place is involved in communication by being referenced by social actors

(Giddens 1979: 206). Landscape has ideological and ontological implications for the way in which one interprets the world (Tilley 1994: 25). "Landscape … reminds us of our position in the scheme of nature" (Cosgrove 1989: 122) thus it can confront the individual's conception of self with that of the other. The dwelling is a state of one's in-depth familiarity with place, and the building involves the transformation of place through the purposeful addition of meaning. Dwelling and building are fixed locations that maintain identities (persons and places) by situating memory. Bender (1999: 35) illuminates that in contrast to the linear narrative of the West, "memory collapses time into space". Thus space in essence becomes a landscape of memory. Emotion is involved in this process as individuals become attached to space and/or respect the landscape (Tarlow 2000: 719). In this sense in a cognitive account landscapes are the reflection of the cultural cognition group, and in a phenomenological account are composed of cultural meanings that may embody personhood. Thus the landscape is often envisioned in anthropomorphic terms.

Landscape is created through activity both physical and conscious. Landscape is produced by movement that connects social relations and practice but also other places. For as Tilley (1994: 31) notes "places are read and experienced in relation to others" thus movement is involved in interconnecting experience. For as the individual can move, so too can the landscape. Strathern notes how the Melanesians make "places travel" (Strathern 1991: 117). Furthermore Islamic traditions are seen to transpose place also (Insoll 2004: personal correspondence). Heidegger (1971a: 157, cited in Thomas 2001: 173) notes that whilst an individual can physically only be in one place at a time their dwelling is realised as pervading a much more extensive area. Hence dwelling concerns directionality and nearness/distancing of positionality. Indeed the landscape is carried around within the cultural subject through memory. Landscape retains its meaning and these place meanings, understandings and obsessions are imposed upon other places. Landscape is thus simply humanised space and thus relational, temporal and specific. Meaning acquired through agency centres space.

> "A centred and meaningful space involves specific sets of linkages between the physical space of the non-humanly created world, semantic states of the body, the mental space of cognition and representation and the space of movement, encounter and interaction between persons and between persons and the non-human environment" (Tilley 1994: 10)

Culture, emotion, memory and landscape are all interrelated. A cognitive notion of the memoryscape is thus an expression of the convergence zone that homogenises these characters. In contrast the phenomenological memoryscape is a refinement of the conceptual *maps of meaning* promoted in the discipline of human geography. This concept champions the idea that cultures are planes of meaning through which individuals sustain intelligibility and comprehensibility. Numerous similarly useful metaphors have been taken up by cultural geographers e.g. 'maps of meaning' (Clarke *et al* 1976; Jackson 1992), the 'geography of the imagination' (Chambers 1986; Davenport 1984), 'sense of place' (Feld and Basso 1996), 'symbolic space' and 'cartography of taste' (Hebdige 1988). It should, of course, be mentioned that maps and other cartographic enterprises have been criticised for the power relations they embody in their mis/representative omissions and inclusions (Thomas 2001: 168).

The concept of social memory has received much recent anthropological attention (see Teski and Climo 1995; Antze and Lambek 1996; Jeffrey and Robbins 1998; Werbner 1998; Archibald 1999; Climo and Cattell 2002a) and some of this scholarship contributes substantially to understandings of the 'externalisation' (that which pervades/permeates the human body) of memory. In the absence of memory "the world would cease to exist in any meaningful way" (Climo and Cattell 2002c: 1). Memory is always found in context and depends on cultural vehicles for expression (Lambek and Antze 1996: xvii). Memory is complex.

> "[M]emory, whether individual or collective, is constructed and reconstructed by the dialectics of remembering and forgetting, shaped by semantic and interpretive frames, and subject to a panoply of distortions (Climo and Cattell 2002c: 1)

Collective memory should not be conceptualised as a supra-ordinate group consciousness but rather shared collective thought created through the interactions of individuals as members of cultural groups (2002c: 2). Collectivities construct their 'representations' of the world through agreed upon articulations of the past, versions reinforced through communicative endeavour (Yerushalmi 1982; Halbwachs 1992). At this level private remembrance features only as performative reinforcement of such versions. Indeed collective memories are expressed in a multiplicity of conducts. Brundage (2000: 5-13, in Climo and Cattell 2002c: 4) assert that collective memories create and sustain interpretive frameworks that are implicated in making experience comprehensible. Collective memories are characterised by the dialectic between historical contingency and stability and innovative change (Brundage 2000: 11). Nevertheless individual memory cannot be ignored. Individual and collective memories are inextricably intertwined in social memory. Individual memory is highly subjective but it is also profoundly social.

> "Memories define our being and our humanity as individuals and in collectivities. Moreover,

the individual consciousness by which we recognize ourselves as persons, and the collective consciousness by which groups identify and organize themselves and act with agency, arise from and are sustained by memory" (Climo and Cattell 2002c: 12)

The relationship between the collective, the individual and memory is debated by numerous academic traditions. The basic tension arises in the theorisation and contemplation of whether life histories are portraits of unique individuals with special characters, capabilities and skills or whether the same person is reflective or representative of wider socio-cultural configurations (2002c: 23).

In both cognitive and phenomenological accounts memory is the retention of, or the capacity to retain and recall, past experiences or previously acquired knowledge or skills, which are bound up within the body, bodily practices (Csordas 1994, 1999; Stoller 1995; Sullivan 1995; Archibald 2002), and the worldly objects and their embodiment in rituals. McLuhan makes vague reference to a West African prince who made sense of his first encounter with a written language by commenting that the "marks on the pages were trapped words" (McLuhan 1964: 81). Landscape in a similar fashion 'traps' memories, thoughts, events, stories, and myths upon itself. These trappings are performed rather than read. Performative memory involves the storage of knowledge in rituals and may also incorporate involvement of objects laden with memory or simply collective/individual acts of remembering. Traditional archaeology's focus on the inscription and marking of the landscape, which is taken as indicative of human agency, has resulted in the meaningfulness of those landscapes that are unmarked being downplayed. People's identities unfold within the social construction of a sense of place (Wilson and David 2002: 1). This place may or may not be physically inscribed but it will undoubtedly be culturally and consciously constructed. People remember communally as well as individually and these shared memories involve cooperative work in often localised settings (Hind 2004: 36; Urry 1996: 50). Memories, like histories, are experienced in a fundamentally futural way in that they are an active approach or setting for the past which informs and relates to personal and/or collective aspirations for the present and the future. All landscapes embody memories. Moreover "through mnemonics the past is continuously drawn into the present as identities are crafted" (Wilson and David 2002: 6). Although clearly representationalist in intellectual loyalty Casey (1987: ix) sums this up as follows:

> "[E]very fiber of our bodies, every cell of our brain, holds memories – as does everything physical outside bodies and brains, even those inanimate objects that bear the marks of their

past upon them in mute profusion … [memories] take us into the environing world as well as into individual lives"

What are the processes of social memory involved in maintaining/constructing meaning from the past? Social memory is continuous and cumulative yet simultaneously provisional and contingent (Climo and Cattell 2002c: 25). Can processes harmonise these observations? Schudson (1995) describes the processes of *distanciation*, the weakening of intensity and detail over time, *instrumentalisation*, the influence of the present on the past and the past in the present, *narrativisation*, the encapsulation of memory within cultural vehicles, and *cognitivisation* and *conventionalisation*, making the past knowable through conventions such as autobiographies and monuments. Furthermore the processes of re-membering (Myerhoff 1992), personal event memory (Pillemer 1998), unremembering and forgetting (Makoni 1998), co-remembering, re-remembering (Archibald 2002), sociobiographic memory (Olick and Robbins 1998) and the sedimentation of memory (Schutz 1972) have also been proposed. These memory processes "overlap, interact, and … do not confine themselves to rigid categories" (Climo and Cattell 2002c: 25). Indeed these processes are simply characteristics of memory and consequently apply to all memoryscapes. In contrast Meskell (2003: 34) proposes the disentanglement of commemorative practice, performance and short-term memory, with cultural memory, long term projects displaying hybridity through changing temporalities. Such disentanglement might be beyond the 'archaeologies of memory' (see Van Dyke and Alcock 1995: 9). In reality such a project of disentanglement is a futile endeavour considering that even if such categories of memory do exist both strands are so inextricably enmeshed that they can only exist whilst inhabiting a dialectical relationship recursively shaping each other.

What the properties of the environment mean to an individual is a function of its own being and its orientation to the world which confers a scheme of perception (Hopkinson and White 2005: 20). This scheme is an appropriation or *project for living* (Ingold 1986). The subject renders the world a landscape through its engagement with it. The environment is defined by the subject and for some part emanates from within them (Rose *et al* 1984; Hopkinson and White 2005). The memoryscape thus might be conceptualised as structural in the sense of environmental coherences and rules. These structures are lived and, as has been proposed in the context of the Aboriginal Dreaming, "allow the evocation of connotations through their incorporation in subsequent history, accommodating the exigencies of successive historical events" (Hind 2004: 47). A syncretistic dynamic exists where the new is made intelligible through the traditions and resonances of the past. In this sense memoryscapes relate to dwellings. Memories, again like histories and pasts, offer rootedness

and attachment. Thus temporal fixity and being-somewhere results in the feeling of belonging. This sense of belonging or attachment is culturally enhanced and might be manifest in the human preoccupation with boundaries, home-ranges and ethnicities. Undoubtedly the memoryscape is a manifestation of an innovative 'style of thinking' that pervades all aspects of the experiential worldview. It has been repeatedly noted that mentally Africans bring their ancestors with them when they move regardless of where they are buried (Kopytoff 1987; Odner 2004). It is perhaps foolhardy of these authors to think in terms of a pan-African tradition. Nonetheless this is a recurrent theme within Africanist ethnography and archaeology. In a similar vein one can propose that all humans take their memoryscapes with them when they move regardless of where they originate. These memoryscapes are supplemented with new environs but the landscape or background travels.

Landscapes are the product of culture manufactured and shaped through occupation and perception (Schama 1996: 9-10). Just as cultural meaning is performed through a habitus landscape is understood in terms of tradition. This is clear when the endurance of landscape myths and memories are considered. Western culture has become accustomed to separating nature and human perception into two spheres. This is erroneous as the landscape is a product of being. Landscape is "built up as much from strata of memory as from layers of rock" (1996: 16). The landscape is a codification of history from the viewpoint of personal and group expression (Stewart and Strathern 2003: 1). Although one should conceptualise landscape as a cultural embodiment rather than as an *aide-mémoire* of significant inscribed human actions (Küchler 1993: 85). It is a component of the template of memory activity – landscape as memory process rather than memory inscription (1993: 86). Küchler's (1993: 87-100) exploration of the landscape embodiment of New Ireland, Papua New Guinea highlights the processes of "forgetting of place", "handing over to forgetting" and "recollecting the forgotten". In this context it is useful to note the notion of the mental landscape map (1993: 96) that retains memories, myths, and contexts concerning past, present and future agency for recall. These templates relate and entwine memory and landscape together. Thus the memoryscape is composed of systems of signification embodying, conveying and retaining multiple – often conflicting – discursive fields. These discursive fields contain meaning, contexts, symbol and embodiment and myths (Cosgrove 1993).

"Myths themselves constitute discursive fields or narratives purporting to represent specific human experience, but resonating across time and space. Myths may both shape and be shaped by landscapes, not only by those localised and specific landscapes visible on the ground, but equally by archetypal landscapes imaginatively constituted from human experiences in the material world and represented in spoken and written words, poetry, painting, theatre or film" (1993: 281-2)

RELIGIOUS MEMORYSCAPES

The impression of the memoryscape as frequently being religious, or as embodying the sacred, is an extension of the preceding theorisation associated with the cultural memoryscape. If landscape is recognised as place connected to a cultural past (Schama 1996: 16) it is meaningful and evokes individual and collective memories. Landscape therefore is the product of perception. Others note that totemic relations involve connections between persons and places (Ingold 2000: 113). Places embody the essential ties between all beings (Fowler 2004: 122). Furthermore the ancestry of humans and animals can potentially be traced from place and hence contain energies and potencies (Ingold 2000: 113). It has recently been proposed that in certain historical contexts animals, places and artefacts may have been considered sentient beings who played active roles in the social world (Brück 2001b; Fowler 2001, 2002, 2004; Conneller 2004; Pollard 2004). These proposals are grounded in theorisation of personhood that suggests the boundaries between people, animals, artefacts and places can be collapsed. In these cases things may have been deemed constitutive of personal and collective identity and thus the boundaries of the human body were extended. This sense of personified surroundings is one mechanism that may make sense of animistic conceptualisations of the world. In relation to these processes is the notion of *genius loci* or 'spirit of the place'. This conception is particularly important to the theorising of the memoryscape. Indeed it is influentially noted that the "cultural habits of humanity have always made room for the sacredness of nature" (Schama 1996: 18). Landscape is invested with the "inescapable obsessions" of humankind (1996: 18). Landscape is inscribed with memory that is made intelligible through the appreciation of a sense of place. The memoryscape is dynamic and fluid. The attachments and meanings found in the memoryscape can be transformed and subverted having been produced through the medium of culture (1996: 32).

> "[L]andscape tradition is the product of shared culture, it is by the same token a tradition built from a rich deposit of myths, memories, and obsessions. The cults which we are told to seek in other native cultures – of the primitive forest, of the river of life, of the sacred mountain – are in fact alive and well and all about us if only we know where to look for them" (Schama 1996: 14)

People develop emotional attachments to places (Climo and Cattell 2002c: 21). Meaning resides in these places

and thus they have been cast as *memory places* (Archibald 2002) or *landscape memory* (Schama 1996), as well as the term favoured here – *memoryscapes* (see also Nuttall and Coetzee 1998). Climo and Cattell (2002c: 21) cite numerous useful examples of studies that relate memory and meaning to place e.g. in the natural world (Boyd 2001; Trimble 1995), in created landscapes (Korp 2000), in urban spaces, and in sites of human occupation (Feld and Basso 1996; Zeleza and Kalipeni 1999).

The primary role of memory in preserving, transmitting and negotiating material culture has been stressed (Bradley and Williams 1998; Gosden and Lock 1998). These arguments, as has been mentioned, have also been applied to the relationships between places, spaces and monuments. Such processes include the commemoration of place by monuments (Thomas 1999), the monumentalising of natural places (Bradley 2000), the forgetting and rediscovery of the past (Mullin 2001), and the remembrance of rites (Last 1998). These frameworks characterise the monument as a human response to landscape. Hence the key to archaeologically appreciating memoryscapes and elements therein is the notion that the place, feature or monument acts as a memorial whose significance is obvious, although probably no longer knowable. It has been argued that memories of events that occurred at a certain site may have been vital to group identity and that monument construction may have been a single, and certainly not exclusive, potential response to such events. Mullin (2001: 535), for example, has proposed that storytelling may have been another means of ensuring the longevity of memories.

The production of the memoryscape is a compulsive characteristic of human potential inherent within the act of dwelling/worlding. In the cognitive tradition the link between human memory function and the universal characters of religion has been forwarded in the *Two Modes of Religiosity Model* (Whitehouse 1995). In this model imagistic modes of religious practice rely on episodic/autobiographical memory whilst doctrinal modes are grounded in the cognitive constraints of semantic memory. The point should be highlighted that the religious tendencies precipitated by these memory systems are hugely significant in the construction of the memoryscape. Consequently cultural memoryscapes are likely to exhibit considerable degrees of overlap and conformity to either the imagistic or doctrinal modes due to both the commonalities of experience of cognition groups and the mutual interdependence of certain characteristic corollaries of the religious modes. Other theorists have come to related and similar conclusions through vastly differing methodological and epistemological perspectives. Jung, for example, believed that the universality of nature myths testified to their indispensability in appeasing the interior psychological traumas by ordering the world. Further to this Eliade argued for nature's full operational presence in all contemporary cultures. This is because some religious beliefs are optimally conceptualised as mythic. Thus they are not believed to be literal truths but symbolic ones that express basic feelings and notions about the world within traditions of interpretation (Argyle 2000: 91-2). Religious knowing is rendered intelligible through the mutual discovery, rather than invention, of expressive symbols by individuals and societies (Watt and Williams 1988). Within this framework emotional attunements have primacy. Emotional responsiveness is the religious mode of knowing especially in those popularly understood as aesthetic appreciation: awe, marvel, reverence and empathy (see Watts 1996). Sensation has been posited as being associated with religious landscapes. Bradley (2000: 32) notes "experiences rooted in nothing more arcane than the nervous system" whilst discussing shamanic sensation, religious revelation and particular kinds of location. Indeed *An Archaeology of Natural Places* (Bradley 2000) explores case studies that link landscape and sacred place, and by implication asserts elements of religiosity found at those locales.

Stewart and Strathern (2003: 3) highlight the embedded sense of place within mythical, ritual and local landscapes. These senses of place serve as pegs upon which individuals and communities hang their memories, derive meaning, and establish religious arenas of action. These senses of place thus serve as markers of present and projected identity as well as continuity with the past. Places should not be considered as mere fixed points upon the landscape but should also incorporate the sense of fluidity associated with movement through points (see Rapport and Dawson 1998). Thus the theories of belonging and attachments to landscapes should be extended to incorporate the travelling of inner landscapes. This process has been categorised as the 'inner landscape of the mind' where knowledge is transportable. Knowledge can remain private or become objectifiable through sharing (Stewart and Strathern 2003: 7). This, of course, highlights the subjectivity of emplacement. Pivotal to theorising landscapes is thus the appreciation that "the intersection of memories and history with how people see themselves in relation to their environment is vital to the understanding of people's placements and movements through social contexts" (2003: 10). In order to further pin down some religious dimensions of certain memoryscapes one needs to define and consider the themes of immanence and animism.

Immanence concerns the notion that the sacred is very much associated with creation, is all-present in the world, and is close to all humankind. Immanence thus expresses the way in which the sacred is thought to be present in the world. The most extreme form is pantheism, which identifies divine substance either partly or wholly with the world. The most diluted/weakest form of immanence is expressed in eighteenth- and nineteenth-century deism, in which god creates the world and institutes its universal laws, but exercises no providential activity and is basically an 'absentee landlord'. Clearly then immanence is not directly concerned with the supernatural. In contrast the supernatural is more closely allied to transcendence. Transcendence generally means 'of the deity' but includes the attribute of being above and independent of the universe; a state of being or existence above and beyond the limits of material experience. A transcendental being is incomparably superior to other things. The supernatural has to be considered though because it is so frequently associated with the sacred and the way it has been academically conceptualised. If the sacred is immanent in the physical world religious experience does not have to be about transcendence at all.

Animism is linked to immanent worldviews in that it recognises that the world is full of persons, only some of whom are human. The term has been used in religious discourses for quite some time and upon close inspection it is evident that animism means different things to different theorists. The term clearly began as an "expression of a nest of insulting approaches to indigenous peoples and to the earliest putatively religious humans" (Harvey 2005: xiii). In contrast it has been proposed that with appropriate vigilance and regular refinement the term might be beneficially used to study religious experience (2005: xiii). Quite simply the older usages of the term constructed animists as backward people who refused or were incapable of distinguishing between objects-subjects or inanimate-animate. More recent conceptualisations of animism, in contrast, theorise worldviews and lifeways in which people seek to know how they can recognise and respectfully engage with other persons or immanent beings. The older uses of the term 'animist' were used by academics, colonialists and missionaries alike to present a primal religious substratum onto which more advanced religious appreciations could be imposed. Even if the inherent racism of the colonial project is ignored such theorisation has been demonstrated as being fallacy for one cannot 'add-on' characters to an incomplete primordial body.

The term animism has had a long and contentious intellectual history and it would be prudent to review some of its former uses and associations. The term itself was derived from early 18th-century chemistry and the theory that the physical element *anima* vitalises living bodies whilst its reciprocal element *phlogiston* enables some materials to change form (Stahl 1708). Hume expressed animistic notions in his religious philosophy during the mid-seventeenth century noting that "there is a universal tendency amongst mankind to conceive of all beings like themselves" (Hume 1956: 3). Thus humans attribute the characters of humanness to the world. Similarly, others in the later nineteenth century continued linking animism with primordial religious perspectives (Fraser 1976: 146; Tylor 1924: 500). Importantly animism was conceived as a human attempt to make sense of the world. This is potentially useful but these accounts are flawed in their underdeveloped references to 'souls' and 'spiritual beings' and their reliance on data amassed through the colonialist enterprise. Durkheim (2000) proposed that totemism was the original religious system noting that social forces such as kinship related sub-communities to a symbolic animal or totem. These social relations were later expressed in cultural regulation and prohibition. More recently, animism has been argued as being natural to humans and other species (Guthrie 1993, 2001). In this account animism is a predisposition rather than a conscious act and has been positively selected for throughout evolution, because of the associated survival advantages of considering things and places to be capable of hostility rather than not. It is important for Guthrie (1993: 39-40) to distinguish between the predisposition to animate (attribute life to) and anthropomorphise (attribute humanness to). This theory has been severely contested for not actually showing why what is little more than a survival instinct should be so central to religious experience (see Bird-David 1999). It should also be mentioned that panpsychism is another form of animism. This philosophical stance recognises mind, experience, sentience or consciousness in matter. In panpsychism all objects possess an interior, subjective reality i.e. they are subjects (de Quincey 2002: 104). There is something it feels like from in a body i.e. subjectivity that "goes all the way down in all forms of matter" (Harvey 2005: 17). The new animism is distinct from these 'animisms' and theorises worldviews and lifeways in which people seek to know how they can recognise and respectfully engage with other non-human subjects (see Harvey 2005).

The new animism is a diverse investigation into ways of living that are life-affirming and respectful. It strives to provide the means to speak, listen and interact with a world full of cacophonous agencies (Haraway 1992: 297). It is established on indigenous and non-modern ways of being. It explicitly acknowledges that many indigenous phrases may be translated as 'the good life' or 'living well' and most of these are linked to ways of behaving and centre on the inculcation of respect and/or deference. In this sense the new animism celebrates that humans are all sensual, embodied, participative persons in a sensuous, relational world. The new animism notes that there are, and always have been, multiple ways of living respectfully in the world but that the processes of modernity have diminished our appreciations of the world. Indeed there are many reasons why humans have found themselves unable to relate to the modern world. Firstly, writing has separated and abstracted the sensuous world from sensuous bodies because visual symbolism is given primacy (Harvey 2005: 206). Secondly, the spread of mass agriculture has blinkered humans from seeing many of the natural cycles of growth. Thirdly, the displacement of humans from the environment in the commercialisation, urbanisation and industrialisation of the world has removed them and obstructed relationships forming between them. Fourthly, the Cartesian project has achieved a 'schizophrenic split' between mind and world and proposed that humans are surrounded by nature and resources which inevitably leads to exploitation (2005: 206). In this regard it is noted, however, that the ideologies of modernity have never fully succeeded in separating and reclassifying the animate world. Harvey (2005: 207) notes, despite genocidal and ecocidal attempts to obliterate alternate ways of understanding what it means to be human bodies and surroundings, that such understandings are resistant to even the most extreme forms of persuasion. Humans have too many experiences of 'alivenesses' in the world that the world will always (at least in part) be celebrated as animate. Thus it should be recognised that some writers, some of those who do not grow their own food, or some of those attuned to the Cartesian worldly frame will still recognise some animation of the world.

The memoryscape is characterised by immanence. The world is animated. Essentially in light of this theorisation the cognitive scientist would argue that the human mind is preconditioned to personify the inanimate world and have teleological thoughts concerning the origins of that world. In contrast a phenomenologist might argue that if one fully gives themselves over to elements of the memoryscape then they can invoke a response related to or part of the sacred. The world becomes religiously significant to humankind through acts of remembrance. Traditional religions animate the world and can be characterised as connecting people to the forces of nature through the celebration of the sacred (Hayden 2003: 9). Similarly Durkheim's (2000) sociological approach strived to establish the nature of the relationship between religion and society. Durkheim saw religion as a social truth, a collective representation which was shared by a society. Religion consisted of rites and beliefs which centred on maintaining the essential separation between two basic categories, sacred and profane. Sacred and profane corresponded to a basic division between the soul and the body and the individual and society. Religion was essentially a symbolic system by which society could be represented to itself. Durkheim thought that religion, as a representation of society, was continually recreated in ritual. Consequently all religions are factual in their own fashion. Therefore if religion is a reflection of the society that produced it then this supports assertions that all societies have memoryscapes, and that sacred place features in all religions albeit in some cases in a purely metaphorical fashion. The dualism sacred/profane is problematic as typically the sacred permeates all areas of experience but the notion of societal representation is offered limited support by studies that propose landscape features in the experience of religiosity. Awareness of a sacred presence in nature was reported by 16% of respondents in one study (Hay and Heald 1987). In another survey 49% of the study group reported that they had experienced the beauty of nature in a deeply moving way that had had a considerable lasting effect on their lives and a further 33% reported that they had had such experiences but without the lasting effects (Wuthnow

1978). Moreover 45% of those surveyed in another investigation reported the beauties of nature as being the triggers for their religious experience (Greeley 1975).

That the memoryscape features in numerous religions is indisputable. Obviously it would be academic fallaciousness to suggest that the memoryscape exclusively represents religion. There are countless other components, themes and motifs to religion. Nevertheless by holding religions in one investigative field the memoryscape is identifiable throughout. As Argyle (2000: 48-9) notes "[a]ll religions have sacred places, and places may become sacred because they have striking features". Such sentiments that relate what is culturally considered 'striking landscape' to sacred space are frequently echoed.

> "Over and over again these sacred places are connected with, or are, what the western world classes as 'natural' features of the 'landscape', such as mountain peaks, springs, rivers, woods and caves" (Carmichael et al 1994: 1)

Religiousness and sacredness is not uniformly distributed across the landscape but is concentrated upon certain features – thus one can think in terms of degrees of sacredness (see Hubert 1994: 16-8). This does not imply that entire landscapes cannot be sacred. As Chief Seattle famously claimed on the signing of the Treaty of Medicine Creek in 1854 entire landscapes can be ritually significant

> "Every part of this country is sacred to my people. Every hillside, every valley, every plain and grove has been hallowed by some fond memory or some sad experience of my tribe" (Chief Seattle, quoted in Turner 1989: 192; Hubert 1994: 16)

The appreciation of the immanence of the memoryscape elucidates how sacredness can be found throughout entire landscapes or componentially at specific places or features within landscapes. The sacred is part of the context of engagement that is composed of histories, traditions, memories and myths and much more, and thus the landscape is constituted by the pre-understandings of the disclosing-subject.

Religions have culturally evolved, in an organic sense, into complex systems. The memoryscape is a primordial form of what cognitive archaeologists would label external symbolic storage (Renfrew and Scarre 1998). The phenomenological archaeologist would propose that meaning is inscribed or found through the engagement of the environment. Individuals draw meaning from the landscape and ascribe particular places with specific and sometimes contradictory meanings. Thus in documenting the memoryscape one should employ methodologies from 'ethnogeography' that relate environmental and

social settings (Theodoratus and LaPena 1994). It would be prudent to investigate whether the notion of the memoryscape, including the suggested religious connotations, sits comfortably with some of the more traditional discourse concerning religion. Such practical sentiments have been recently rearticulated in other attempts to engage religion with newfound academic rigour. Andresen (2001: 276) advises that the classical theories of religion need to be revisited and explored in light of recent scholarship to consider whether insights have been overlooked. Indeed it will be shown that some of the theorising produced by the 'classic' scholar of religion can be usefully conflated with the notion that elements of religion are subsumed within the memoryscape. This is not to suppose such classic theorisation should be agreed with wholesale but rather critically re-examined.

MEMORYSCAPES AND THE CLASSICAL THEORIES OF RELIGION

Mircea Eliade was prolific in his writings within the scholarly genre history of religions. His work has had considerable popular and academic impact (see Allen 2002: xi-xviii for overview). Importantly one must be wary of Eliade's often beguiling theories. He has been characterised as "one of the more popular examples of the armchair scholar in the ivory tower of the university" (Kehoe 2000: 40). This relates to his tendency to produce sweeping syntheses of conforming evidence whilst making similarly sweeping omissions of evidence failing to correlate. Moreover he seldom undertook any fieldtrips to the locations and case studies he wrote about (see Dudley 1977; Gill 1998; Kehoe 2000). Thus his theories must be utilised with responsibility and questioned at all levels. This does not mean that all his scholarly work is ineffective and misleading but rather nothing can be assumed to be of sound method and interpretation. This is particularly important as it will be demonstrated that some of his insights are fruitfully conflated with the proposed conceptualisation of the religious memoryscape. So what Eliadean themes are usefully integrated? Essentially the sacred is the focus of religious experience (Eliade 1960: 17). Religious experience is the engagement with kratophanies, hierophanies and theophanies, which oblige the recognition of the dichotomisation between sacred and profane (1960: 17). The kratophany is a manifestation of power (Eliade 1958: 14), the hierophany is the act of the manifestation of the sacred world which cross-culturally applies to all religious experience (Eliade 1957: 11), and the theophany is an appearance of God or the divine to humankind. These notions of hierophanies, kratophanies and theophanies are dubious in their ambiguity and relate one can argue to an inability to adequately conceptualise the sacred. As previously noted issue must be taken with the binary opposition sacred/profane. It is, however, unquestionable that one can assert landscape can never be experienced as purely natural phenomena.

"For those who have a religious experience all nature is capable of revealing itself as cosmic sacrality. The cosmos in its entirety becomes a hierophany. The same experience is one of re-entering the mythical paradise of the beginnings" (Eliade 1957: 12)

Eliade proposed the religious commonality that all religious individuals have a "desire for being" (Saliba 1976: 64). The religious strive to remain in close proximity to the sacred order. This objective is realised through inhabiting a sacred universe i.e. *Homo religiosus* makes the landscape sacred by investing or finding religious meaning. Nature thus always has religious value and the sacred becomes bound up with the natural. Thus "he is led to the former by reflecting on the latter" (1976: 65). The landscape is elucidated in terms of divine activity. Thus the world "presents a supernatural valence" (Eliade 1957: 138). Religions enable one to harmonise oneself within the rhythms of nature (Allen 2002: 102). The individual becomes integrated as part of a meaningful cosmic landscape. Eliade situates religious phenomena within a hermeneutic framework of coherent systems of symbolic associations (2002: 103). Intelligibility is attained through fixing the phenomenon within a structured whole. Places become meaningful through their symbolic relation within solar, aquatic, vegetative, faunal, and meteorological cycles. Thus Eliade's method highlights a process of cosmicisation (2002: 103).

Sacred space becomes value-laden as it takes on importance for the individual. Space is contrasted for meaning establishing the binary opposition sacred/profane (Eliade 1957: 28). In reality all space may be conceptualised as sacred, there are simply differing degrees of sacredness that may relate to level or proximity of immanent forces. These degrees are influenced by all manner of variables including time, cultural attunements, the individual, affiliations and rituals. The sacred is non-uniformly distributed upon landscape and certain features will be invested with more meaning than others. Eliade (1957: 88-9) notes that religious individuals continually locate themselves within the sacred past – reading religious meaning from the landscape (Saliba 1976: 64). The sacred landscape therefore is the dwelling of being. The primordial physical landscape becomes value-laden through contrast and reference to the other.

"Religious man assumes a particular and characteristic mode of existence in the world and, despite the great number of historic-religious forms, this characteristic form is always recognizable. Whatever the historical context in which he is placed, *homo religiosus* always believes that there is an absolute reality, *the sacred*, which transcends this world but manifests itself in this world,

thereby sanctifying it and making it real" (Eliade 1957: 202)

The landscape is a living cosmos that is articulate and meaningful (Allen 2002: 104). Human beings comprehend their mode of being-in-the-world when they intuit or decipher meaning from the landscape. Nature is thus an inexhaustible series of Eliadean ciphers that communicate their inherent meaning to human beings (2002: 105). Landscape is narrative. When one intuits fundamental truths one has engaged the memoryscape. Thus landscape is both the medium and the outcome of the memoryscape.

"The World 'speaks' to man, and to understand its language he needs only to know the myths and decipher the symbols...The World is no longer an opaque mass of objects arbitrarily thrown together, it is a living Cosmos, articulated and meaningful. In the last analysis, *the World reveals itself as language*. It speaks to man through its own mode of being, through its structures and its rhythms" (Eliade 1963: 141, emphasis in original)

Landscape will always have creation meaning attached to it that usually has a religious character. It presents itself as the "work of the gods" (Eliade 1957: 117). It is clear that certain places within the landscape reveal greater sacredness than others. Eliade proposes the notion of the celestial sacred to account for the numerous religions that equate the transcendental and remote nature of the sky with sacred phenomena (see 1957: 118-21). This emphasis is misplaced. The skyscape is temporary in nature. The landscape, in contrast, due to its longevity and permanency permits the sometimes protracted processes involved in engaging or attributing the memoryscape. Nonetheless, the skyscape and the seascape are components of the wider memoryscape. Religious meaning found in landscape is disproportionate. It has been mentioned that what are subjectively considered striking landscape features tend to have more meaning intuited from them. The divine is "manifested in the different modalities of the sacred in the very structure of the world and of cosmic phenomenon" (1957: 116). Eliade proposes that the mountain is an *axis mundi* connecting the earth with another existential plane (1957: 38). The mountain in a sense aspires and hence "marks the highest point in the world" (1957: 38). These features are 'high' places both literally and metaphorically. The "most high" is a dimension belonging to sacred forces (1957: 119). The mountain 'speaks' to the human being of highness, infinity, eternity and power because of its very presence. It will be highlighted through primary ethnographic research that such conceptions of sacred landscape are observed. Furthermore this research also substantiates other claims made concerning mountains and dwellings.

Eliade conceptualises religion as a human response to the sacred through which the actual human condition is escaped. The religious individual becomes aware of another mode of existence outside of "historical time" (Saliba 1976: 136). Indeed rituals, rather than providing simply an escape from the 'this-worldly' plane, direct the control of the human condition. It is also worthy of note that rituals performed upon the landscape invent and augment the meaning of places within the memoryscape. Eliade's notion of nature perceives it inscribed with a romantic and neutral character (Allen 2002: 105). There are some similarities between Eliade and Heidegger in their appreciation and conceptualisation of nature. Heidegger generally considers nature as something assimilated into culture of being present-at-hand and having practical uses (Polt 1999: 137). Nevertheless he sometimes reveals another stance on the nature of *earth*. Earth shelters the beings that arise from it, it is the foundation of the world and the medium of dwelling, and it offers spontaneity and conceals itself (Heidegger 1993b: 168-74). The concept of earth denotes a more profound way of relating to nature – as something not only to be respected but as something that exists resisting and preceding human manipulation and interpretation (Polt 1999: 137). Moreover Heidegger writes of "the nature…which assails us and enthrals us as landscape" (Heidegger 1962: 100).

> "[C]ulture arises from nature, and tries to understand that from which it arises. Since a culture sheds light on people and their surroundings, it is intrinsically opposed to obscurity and tries to illuminate nature. But…nature loves to hide: there are always limits to what we can understand, and nature tends to reassert itself in its mysterious power" (Polt 1999: 138)

Truth is found in sheltering – clearing and unconcealment are grounded in particular beings (1999: 149). These beings, of which earth is one and art another, shelter truth.

> "[W]e may glance at a mountain and naively assume that it is just 'there', an object that is given to us. Being and truth are then dimmed down. But if we allow the mountain to shelter the truth of Being, we can experience its 'thereness' more fully. We will acknowledge all the ways in which the mountain makes a difference in our world" (Polt 1999: 149)

The mountain, and other places within the landscape, gathers the world and shelter the truth. They are sites that display various dimensions of significance. This significance is the mystery through which the earth and world are presenced (1999: 150). Cosgrove (1993: 285) asserts that "[b]ecoming is inscribed in landscape". In actuality it is Being that is embodied by landscape through disclosure bought about by the subject and by

memory. It is interesting that Heidegger saw the world as becoming a wasteland, noting indifference, the reduction to unbeings, and the oblivion of Being. Therefore true being and care is of historical possibility rather than a universal feature of humanness (1999: 150). This mirrors somewhat the anti-religiosity of the contemporary West. Perhaps the failure to elucidate and disclose meaning inherent in the landscape, with preference given to scientific quantification and sterile description of the background, is a factor in this process of unbeing for *Homo religiosus*.

There have recently been efforts to reform cultural geography with a focus on environmental hermeneutics (see Seamon and Mugerauer 2000). The emergent research agenda draws heavily from Heideggerian phenomenology and some of its rubric is most enlightening. Environmental hermeneutics concern the interpretation of human-cultural modes of dwelling, the historical features of the environment, how experience and technology impact upon the perception of the environment, and the relation between language and environment. From the outset it is vital that landscapes and environments are conceptualised through their perceptive engagement. Essentially one must explicitly acknowledge that agency and its uses inform the intelligibility of the landscape. Thus just as the act of hammering informs one as to the properties of the hammer, ones engagement with the landscape informs us as to its properties. Thus landscapes are perceived differently if they are lived within, travelled through, travelled to, or travelled from (Mugerauer 2000: 54). Landscape is created by signification.

> "We might replace the assumptions of a naïve realism with those of idealism and suppose that external reality is merely a confusing physical stimulus which needs to be ordered and categorized by human consciousness before we have what we call 'environment'" (Mugerauer 2000: 56)

Environment in this sense is synonymous with landscape. Although one can take issue with the notion of an eternal reality, confusing or otherwise, pre-existing human subjectivity the acknowledgement of environment as a humanised focus is useful. Moreover human consciousness is imaginative and creative.

> "[T]he effort of the consciousness to constitute and form environment actually results in misperception or mistakes about what really is. What seems to emerge in fact conceals or obscures environment. In these situations, language and perception fail reality" (2000: 57)

The religious elements found in the memoryscape may be forms of misperception or mistake, to use the phraseology quoted above, but it is the humanising of place that creates the environments in which one dwells. Furthermore it creates a sense of connectedness between places with humans and with other places. This connectivity is both emotional and meaningful (Dovey 2000: 40). The humanly-made intelligible world is always an incomplete and situated one. This worldhood depends on its meaning and significance (Heidegger 1962: 138). Thomas (1993: 29) notes that "we need to recover the 'spaces of resistance', not just the dominated landscape" whilst discussing the totalising and privileging of landscape perspectives. These contested spaces are especially important in religious landscapes where both individual, collective, local and global differences and resistances may be apparent.

Some recent scholarly work that draws on classical theories of religion is also of related interest. It has been noted how mystical experiences are often catalysed by "natural beauty" and these experiences tend to be religious in flavour (Hinde 1999: 187). The attraction of the natural landscape has been posited as "a rather basic matter" that is "due to mechanisms developed early in the evolution of our species" (1999: 187). Hinde cites studies that substantiate these arguments. It has been documented that natural landscapes are preferred to artificial ones. Indeed the viewing of natural landscapes has been demonstrated to reduce stress in traumatised patients (Heerwagen and Orians 1993; Ulrich 1993). Moreover studies have shown that pictures depicting 'peaceful' landscapes are those most commonly displayed in people's homes (Halle 1993). Another landscape feature proposed as being universally attractive is mystery. Mystery frequently contributes to religious experience (Hinde 1999: 188). The attractiveness of landscape-mystery can be translated as the "promise of more information if only one could enter the scene" (1999: 188). This is relatable to the human need to explore, understand (Kaplan 1992) and decipher meaning. Moreover, as noted earlier, it has recently been proposed, that in various contexts places, animals and artefacts are considered sentient subjects who played dynamic roles in the social world.

Some classic theories of religion clearly journey some way toward the conceptualisation of the religious memoryscape and can be productively supplemented with insights from more recent scholarship. The mechanisms that foster the aesthetic appreciation/attraction of landscapes correspond in part to the lived processes that construct the memoryscape. It must not be overlooked in this context that fashions in landscape preferences do change under influences such as time (Thomas 1983). Nevertheless the universal human capability that recognises the inherent animation and sacred character of the environment might always have been apparent. Interestingly Eliade himself hinted at the notion that religion is a cognitive event tied to landscape. He perceived similar arguments writing that religious phenomenon should be set "in its true relation to other things of the mind" (Eliade 1958: xvii). This brings the two 'epistemologically parallel' descriptions of the memoryscape into focus again. On the one hand it is a cognitive representation and on the other it is a component of the phenomenological background.

Recent ethnographic research can be used to highlight the characters of the memoryscape. This case study material demonstrates how fundamental memories are in the production of the landscape and the subject's experience of it. The findings of the study illustrate how little of living within a memoryscape is archaeologically knowable. Thus landscape archaeology must address such insights by either considering the implications of the unknowable upon whatever is knowable or ensuring that any interpretive exercise explicitly recognises the conditions of such interpretation. This is not to suggest that interpretation should be avoided but rather, on the contrary, that the significance of the unknowable be observed and infused within the interpretive exercise. There is, of course, no methodology for conducting such interpretation. As is always the case in archaeology interpretation is reliant upon context and preservation. Essentially the experiential envelope needs to be widened to account for as much as possible concerning the condition of past humanness and the human past.

CHAPTER THREE

BEING-SOMEWHERE: MYTHIC MEMORYSCAPES OF THE WACHAGGA

SETTING THE SCENE: WACHAGGA OF KILIMANJARO

Various contemporary memoryscapes of the Wachagga can be described based upon primary ethnographic fieldwork. The methodology, documentation, some discussions of project execution and additional detail of the fieldwork element of the reseach are provided in appendices. Kilimanjaro lies in the Northeast of Tanzania near the international border with Kenya and is the traditional home of the Wachagga (see **figure 1**). The lower slopes of Mount Kilimanjaro are also referred to as Chaggaland or *Uchagga*. The core area of *Uchagga* is less than 165 square miles (Tagseth 2003: 14). The population of Kilimanjaro district according to the recent census numbers well over 1.3 million (Central Census Office 2003: 45). It is estimated that about 850 thousand of these would consider themselves Wachagga (Wimmelbücker 2003: personal correspondence). The primary economic activity of the area is farming but land shortages have forced changes to the local economic structure and many Wachagga are now wage earners in local urban centres. The Kilimanjaro region has suffered considerably under the AIDS pandemic. Indeed the region ranks third in Tanzania for HIV infection and AIDS-related deaths (Kerner 1995: 123; see also Hasu 1999; Setel 1999 for extensive considerations of these issues).

The exploration of memoryscapes and religious/spiritual meaning invested in or found through landscape has been partially conducted in much ethnography. Although not explicitly recognised as such the phenomenon of the memoryscape has been most fully documented in the studies concerned with the Australian Aborigines (e.g. Bell 1980; Eliade 1973; Head 2000; Morphy 1993; Myers 1991; Stanner 1966; Strang 1997). Similarly work has been produced centred on European (e.g. Graves-Brown *et al* 1996; Schama 1996), North American (e.g. Clifford 1988; Schama 1996), South American (e.g. Corr 2003), and South Pacific (e.g. Strathern 1991) cases. In the Anglophone tradition especially this phenomenon has not been documented for sub-Saharan Africa. Admittedly there have been studies that peripherally deal with the religious memoryscapes of African indigenous collectives but these have not engaged with the subject matter in appropriate detail (e.g. Harris 1950; Thornton 1980). Thus the locating of the study was informed by such appreciation. Furthermore the puzzling lack of phenomenological perspectives in sub-Saharan ethnographies (see Riesman 1986; Mutema 2003)

requires address. In line with the philosophical underpinning of this research the study endeavoured to adopt a descriptive phenomenological direction if the evidence was amenable. The study site was deemed appropriate because in certain contexts it might be considered a 'striking landscape' and it was therefore assumed that the religious dimensions of the memoryscape might be more easily accessed. This does not mean that striking landscapes are the only media of the memoryscape. Indeed quite the opposite is proposed, but what might be considered 'striking landscape' might more favourably assist the reading and sedimentation of such meaning. Therefore far from betraying an essentialist or racist agenda the selection of the site owes to the necessities of establishing a comprehensive depiction of a previously unrecorded memoryscape compiled within an ethnographic framework.

EXPLORING AFRICAN LANDSCAPES: SOME ISSUES

Recent scholarship has produced some outstanding literature on perceptions and perspectives of African landscapes (Cohen and Odhiambo 1989; Werbner and Ranger 1996; Luig and von Oppen 1997; Ranger 1999; Beinart and McGregor 2003). A commonality of these accounts is the understanding that landscapes are made not through any process of sedimentation of history but through the continuous reworking of experience and future potentialities (Luig and von Oppen 1997: 7). It has also been highlighted that one cannot write about a homogenised 'African perception' as perceptions are of course culturally varied and correspansive (see Luig 1995). Memory and identity are inscribed by social practice. Rather than living inside the landscape the normative appreciation of nature in indigenous rural areas is one of appropriation – physically and symbolically. This process incorporates reciprocity and the capacity for shaping according to need and circumstance (Luig and von Oppen 1995). This has been related to the observed appreciative commonalities that enmesh the nature concept with metaphysical/religious connotations. Landscape is a subject that can be worked upon but is never reducible to a mere object (Luig and von Oppen 1995: 20).

Obviously the memoryscape can have a collective character nevertheless it is an individual or positioned form of knowledge. Overlapping spheres of perception are precipitated through culture. Culture in this sense can be usefully envisioned as a technology that facilitates the

Figure 1. Map showing the Kilimanjaro Region in relation to the countries of East Africa

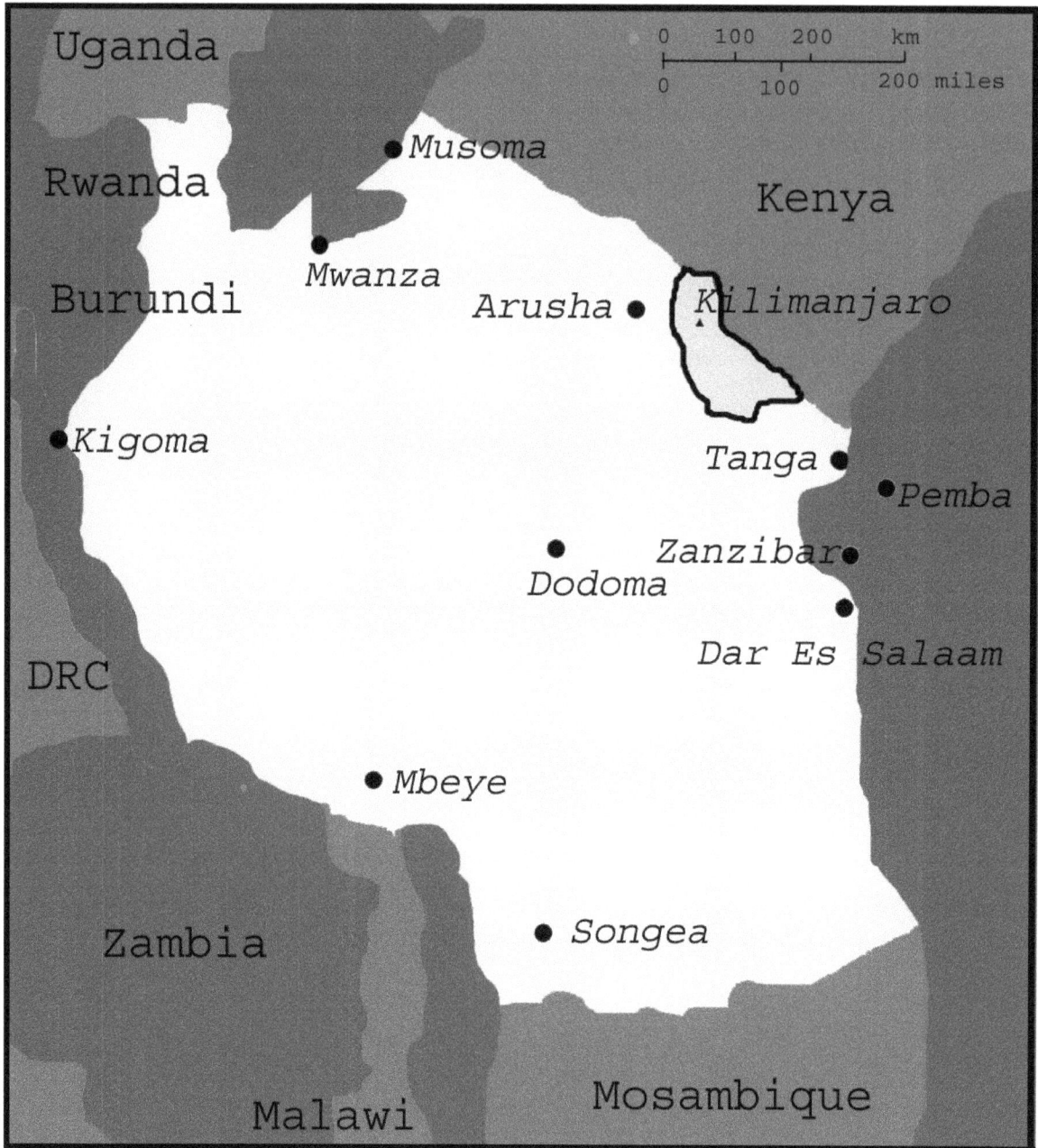

production of meaning through the concernful interactions humans have with their world. Meaning is not fixed within objects and subjects waiting to be located but rather is produced through the workings of the sets of relationships in which humans find themselves (Thomas 1996: 236). Culture is both embodied knowledge and embodied technology. Therefore any research into the memoryscape must be collected from the positioned individual and represented as such i.e. an individual narrative.

Religious conceptions or meaning associated with the landscape are common in African traditional worldviews.

For instance, (Mbiti 1975: 144) reports, "every African people has its own religious places". Furthermore "most of the Bantu believe in natural and local spirits...of mountains and forests, of pools and streams, of trees and other local objects" (Parrinder 1976: 43). The landscape is understood in organic terms "nature is alive in an important sense...many African people worship a god or goddess of the earth and/or sky and believe that spirits are associated with certain trees, hills, lakes and so on" (Mitchell 1977: 55). These observations are frequently reiterated. There are limitless natural places that harbour religious meaning that are symbolic meeting-points between the religious and the non-religious domains

(Mbiti 1975: 146). It seems justified to suggest that the indigenous worldview, informed by the memoryscape, sites landscape in relation to humans rather than in terms of exploitation or conservation (Mitchell 1977: 55). Nevertheless this old-fashioned literature is founded on flawed assumptions such as the existence of an African singularity i.e. the 'Bantu' or a bounded Africa that demonstrates both geographical and cultural continuity. This, of course, is an oversimplification of a highly complex region. Furthermore the religious/non-religious dichotomy reproduces the dualism sacred/profane that has already been documented to be a fallacy. What is worthy of note though is that multiple accounts are in generic agreement concerning how religious meaning is associated with the landscape and environment in various sub-Saharan African contexts.

In the influential paper *Places of Power and Shrines of the Land* (Colson 1997) some pertinent distinctions are raised that concern any conceptualisation of the memoryscape. These build on the insights that natural places of power differ from land shrines (Vansina 1990: 93). The ritual activity associated with nature spirits is conducted where they are believed to dwell and there is usually an absence of land shrines which relate to ancestor spirits (Vansina 1966: 32). Colson distinguishes between places of power and land shrines as ritual sites. The possession of knowledge about these sites act as foci of identity that conveys attachment and belonging to the land. Land shrines contain a relationship to the community as they involve human experience or life-force. Moreover these shrines are deeply rooted in local history and so are embedded in cultural knowledge (Luig and von Oppen 1997: 22). They represent "the continuity of human life forces, not the power inherent in nature" (Colson 1997: 52). In contrast places of power are landscape features that exhibit permanency and are perceived as being inherently sacred – the loci of spiritual power (1997: 48). Potential sites for places of power seem to exhibit surprisingly little variance. Indeed "this is so much the case that few raise the question of why these and not others" (1997: 49). Colson cogently argues that all such sites have the potential to engage imagination and so become imbued with sacred meaning (1997: 49). There is an underlying tautology within the realisation that awe-inspiring objects become associated with strength and force. Obviously both shrines of the land and powerful places exhibit supernatural potency. Although the phenomenon of the memoryscape conflates places of power and land shrines the distinction is still valid and potentially useful. Communities and individuals respond to these sites in differing ways. Religious meaning is constantly re-articulated/re-invented at these locales.

Mountains are particularly good illustrations of striking landscape. Mountains are "natural religious monuments" associated with "beliefs, myths and legends" practiced through "rituals, sacrifices, offerings, and prayers" (Mbiti 1975: 149). Indeed Schama (1996) dedicates an entire chapter to "Vertical Empires, Cerebral Chasms". He writes of the "authentic mountain experience" (Schama 1996: 504), the "scroll of eternity embedded in the rock" (1996: 488), and the mountain as "conceived cerebrally" (1996: 473). Mountains are made intelligible through memory and sensation thus they are potentially emotive and highly stimulating. If some elemental components of religions might be founded in memoryscape it should be feasible to conjecture that religious identities might integrate striking landscape and its inherent meaning into their constitution. The contemporary Wachagga, despite missionary activity and the modernising processes of development, incorporate the environment into the religious worldview.

CHAPTER FOUR

CHOREOGRAPHIES OF RELIGIOUS EXPERIENCE

SYNCRETISM: DEFINITIONAL CONCERNS

As two anthropologists note in their introduction to the influential work *Syncretism/Anti-Syncretism* (C. Stewart and R. Stewart 1994) "'syncretism' is a contentious term" (R. Stewart and C. Stewart 1994: 1). Nonetheless the academy has been enduringly enchanted with the term (Magowan and Gordon 2001: 253). The term is derived in all probability from the Ancient Greek words *syn* 'with' and *krasis* 'mixture' (Stewart and Stewart 1994a: 3) and has been used in Christian theology since the early seventeenth century (van der Veer 1994: 196). The syncretistic notion is frequently utilised to describe the mixture of traditional and world religions especially those brought into contact during the histories of colonialism. Local variations/versions of Christianity are often mechanically labelled as syncretistic (Peel 1968: 140). This is actually a form of anti-syncretism. It is crucial that the category anti-syncretism is accepted for this relates to the processes that precipitate notions of authenticity, purity and primordiality. These political overtones performatively devalue the syncretistic religion as an inferior caricature of the original. Thus anti-syncretistic discourses imply impurity, weakness (Kiernan 1994: 70), confusion and disorder (Mosko 2001: 260). Indeed one commentator stressed the socio-political power of the category by commenting that the syncretistic religion "becomes the bridge over which Africans are brought back to heathenism" (Sundkler 1961: 397). In the anti-syncretistic sense religions are value-laden and variable but always against the notional backdrop of authenticity and purity. The instrumentalist critique of stable primordial traditions (see Greenfield 2001; Hobsbawm and Ranger 1983) has shown such entities to be fallacies. Despite this all religions are convinced of their own orthodoxy and orthopraxy (Stewart 1994: 133).

Syncretism should not be utilised as a categorisation but rather to denote processes and systems of religious synthesis with particular powers, structures and agencies. Thus one should focus upon the discourses of syncretistic workings (Stewart and Stewart 1994a: 7). This focus should recognise that syncretism is distinguished from other forms of cultural change or bricolage in its sole application to religious phenomena (Magowan and Gordon 2001: 255). All religions are syncretistic – synthetically incorporating exogenous cultural, material and ritual elements over time. Religions are creolisations, hybridisations or intercultures of multiple other syncretistic cultural entities and forms (Stewart 1994: 127). The only authentic dimension to religions concerns

the synchronicity of the syncretistic mix (Mosko 2001: 260). Thus syncretistic agendas should seek to inform about modification, assimilation, interpretation and origination (Mosko 2001: 271; Stewart 1994: 128). Moreover the agenda embodies the processes of moral, spiritual and emotional synchronicity (Magowan 2001: 278; Robbins 2004: 314). For the term to be anthropologically employable it must be applied within a relativist framework (van der Veer 1994: 197). Within such a cosmopolitan, pluralist and multicultural framework equal theological and ritual perspectives can be modelled. One could even forward the term 'multisynthesism' to denote such a philosophical adherence.

How was/is religion conceived by African Christians? Strangely there are very few accounts that document how local interpretations of world religions evolve during contact – the processes of appropriation, integration, translation and diabolisation (Meyer 1994: 45, 65). The indigene must be recognised as proactively involved in the shaping of their religious worldview by creatively and imaginatively striving to make sense of religious attunements, teachings and understandings (Magowan 2001: 288). Thus syncretistic processes involve a recasting of indigenous beliefs within a fluid and malleable field of agency, power and value (Magowan and Gordon 2001: 253). It has been noted that ethnographies of the missionary period frequently omit the missionary or coloniser from the project (Pels 1999: 16). The early African assistants and their ministrations are also often peripheralised. This is even more disastrous when the political power of the missionaries is considered (Comaroff and Comaroff 1991: 252-4). These actors are, of course, as central to syncretistic processes as the indigenes are and to underplay their involvement simply misrepresents at best and establishes a reverse discourse at worst. Furthermore it is quite clear that in general the impact of missionary activity was immense throughout the colonial and more recent eras. This deep impact involves indigenisation of ideologies of moral change, ethical codes and institutionalising materials. Therefore syncretistic discourses must and are well placed to acknowledge and theorise wider cultural issues (Robbins 2004: 319). The syncretistic processes of the indigenisation of Christianity were part of a dialectical encounter that involved everyone and everything concerned (Comaroff and Comaroff 1997: 5).

Generally syncretistic processes governing African conversion have been related to the indigenous quest for meaning in the face of modernisation and development

(Comaroff and Comaroff 1991: 249). Essentially identities are subjected to the transformative forces resulting from the incapacity of traditional cosmologies and ideologies to make the world intelligible outside of the local microcosm (Horton 1971: 101; also see Horton 1993 for further discussion). The impact of Christianity, amongst the world religions, was particularly forceful because of its acceptance of an omnipotent and omnipresent deity. Thus the arrival of Christianity was logically interpreted as being orchestrated by the divine. This was part of the indigenisation of Christianity throughout sub-Saharan Africa (Comaroff and Comaroff 1991: 210). Likewise preaching, the means of conjuring the feeling of the numinous through audible proof, and resultant conversions were understood in terms of indigenous knowledge – "the medicine of God's word" (Moffat 1842: 576, in Comaroff and Comaroff 1991: 228; see also Comaroff and Comaroff 1997: 65-73). It would be prudent to examine the syncretistic processes in greater detail through contextual precision.

MISSIONARY ACTIVITY ON KILIMANJARO

Some scholarship has focused on the religious dimensions of experience in tribal-ethnic groups that neighbour *Uchagga* (e.g. Benson 1977; Danielson 1977; Flatt 1980; Harris 1978; Sandgren 1989; Spear 1996, 1997; Spear and Waller 1993) and offer some insight into the local religious milieu. As in these accounts the contemporary religious experience of the Wachagga must be historically contextualised in order to offer any substance and accuracy. The first mission station in *Uchagga* was an outpost of the British Church Missionary Society and was operational from 1885-1892 (see Bennett 1964; Stock 1916). They were evicted by the Germans and were replaced by the Leipzig Mission (Lutheran) and the Holy Ghost Fathers (Roman Catholic). From 1892 until Tanzanian independence in 1961 the Lutheran and Roman Catholic missions were the only groups in Kilimanjaro. At the time of independence over 85% of the Wachagga considered themselves Christian (Iliffe 1979: 467). Even today the efforts of Pentecostal and Islamic projects are having limited conversion effect (Colwell 2000a: 6). From the outset missions were involved in the negotiation of local power relations. Through "strategies of diplomacy, duplicity, and/or collaboration" (2000a: 9) area chiefs used the missions to gain trade opportunity and technological, economic and military advantage. This can be optimally illustrated by the Marangu chief who baptised one son Catholic, another Lutheran whilst remaining pagan himself (Stahl 1964: 325). Missionary outposts brought economic advantage through the introduction of coffee cultivation (Lawuo 1984: 23-4; Ogutu 1972). Chiefs welcomed secular education and some religious teaching (Colwell 2000a: 6; see also Lema 1968, 1973). Moreover contrary to popular demographic assumptions the child mortality rate on Kilimanjaro between 1890-1940, as extrapolated from mission and

church registers, was comparable to that found in European locations (see Colwell 2000a, 2000b). This may simply relate to the introduction of the anti-malarial quinine and other innovative medical practices or perhaps, as others suggest, that the moderate infant morality rates may have been characteristic of the locale prior to missionary presence (Colwell 2000a: 35). The colonial period did witness a phenomenal population increase in the region (Moore 1970: 333). The presence of the missionaries was not utterly uncontested indeed there was some short-lived resistance to some missionary establishments (see Iliffe 1979: 100-2).

The influence of the missions was generally impeded in subtler ways than armed resistance. Influence was culturally contained through the shrewdness and guile of the converts and indigenous authority figures. For instance when the authority of chiefs was challenged or revered cultural practices were discouraged by the missions various tactics were adopted by the indigenous groups to articulate their disapproval. These included interfering with the water supplies of missions, the boycotting of churches, and violent eviction (Colwell 2000a: 6; Meyer 1891: 98; Moore 1977: 14). Nevertheless the missions did have considerable conversion success compared with similar ventures in other areas (see Wright 1976). This can be explained by the underlying philosophies of the missionary schools engaged with the local projects. The Leipzig Lutheran mission possessed an adaptationist rather than assimilative attitude to cultural permeation. Similarly the Holy Ghost Fathers were noted as having adopted policies of Africanisation long before this became a "dominant refrain of Catholic missiology" (Pels 1999: 113). These missionary schools of thought promoted integration, adaptation and acceptance of native customary behaviour (see Winter 1976; Wright 1971). Thus the fortunate situation is apparent where processes of domination and eradication have not been deemed totally applicable. Therefore remnants of the traditional religious practice are still apparent in the religious expression of the local populations. It should also be remembered, in this context, that the wealth of information gathered and subsequently preserved concerning the traditional Wachagga culture owes much to the foresight of the early Leipzig missionaries. The gradual disappearance of old traditions and folktales was posited at the beginning of the twentieth century and measures were taken to document as much information as was practicable (Ittameier 1908: 558).

It has been noted that the dominant missionary enterprise was characteristically of Christian (in particular Lutheran) denomination and German nationality (Hasu 1999: 35). The Lutheran mission on Kilimanjaro has been discussed in some detail elsewhere (see Fiedler 1983, 1996; Smedjebacka 1973; Winter 1979). The missionary enterprise corresponded to the Germanic intellectual tradition of the period. In particular the Counter-Enlightenment notions of *Kultur* and *Volk* are present in the interaction of missionary and indigene (Hasu 1999: 32). Essentially this philosophical stance identified the particular histories of social entities as the only means of understanding human group cohesion (Bunzl 1996: 20-1). Thus *Kultur* was a manifestation of a pre-Romantic national ideological particularism that was diametrically opposed to the contemporary notions of universalism and cosmopolitanism. Thus in contrast to the colonialist enterprises of the French and British, which were philosophically underpinned with the concepts of individualism and utilitarianism manifest in the French Revolution and Western Enlightenment (Winter 1979), the Germanic ideological rationale did not propose the adoption of a standardised condition of civilisation. Nonetheless both humanist enterprises were universalist in their acknowledgement that all humans were potential believers capable of redemption (Comaroff and Comaroff 1997: 65). Moreover common to all colonial missionary endeavours was the visionary objective to extend the boundaries of Christendom throughout the African *terra incognita* (Comaroff and Comaroff 1991: 200). The Germanic missionising agenda was, however, very different to the Anglo-Saxon one (Hasu 1999: 35). Nonetheless considering the clear lack of success the Anglo-Saxon organisations, such as the London Missionary Society and the Wesleyan Methodist Missionary Society, had in instilling their orthodoxies within their missionising project (Comaroff and Comaroff 1997: 7) it seems fair to note that the adaptationist agenda was in reality the only workable one.

It is worth mentioning that Hasu's work is more comprehensive than other ethnographies of the Wachagga in its consideration of local religious subtleties. The work adopts an inclusive approach to the published body of material of the Leipzig Missionary Society *Evangelisch-lutherisches Missionsblatt* (1893-1941), the journal of the Leipzig women's mission *Lydia* (1908-20), and the journal of the Northern Diocese of the Evangelical Lutheran Church of Tanzania *Umoja* (1990-4). This may relate to the work being a self-styled 'historic ethnography' and hence must consider a greater body of sources. This historic ethnography informs that the arrival of the colonial administrators and missionaries was interpreted by the indigenous people "according to their cultural understandings thereby making the encounter into a culturally ordered event" (Hasu 1999: 39). In this sense perhaps syncretistic processes are better described as an indigenisation of Christianity and modernity? Indigenous and external cultural structures, exogenous cultural forms, asymmetric power relations, and forces of resistance, proletarisation and domination compel behaviour and relationships within communities of contact (1999: 40). The cultural repertoire is permeable and inclusive. Cultures subsume foreign subjects into logically coherent correspondence (Sahlins 1993: 15). Syncretistic process involves the performance of categories and subjects. Categorisation results in

codification. From the 1920s the categories *kizungu* (white/European) and *kikristo* (Christian) emerged as common categories indicative of wealth, respect and currency. Furthermore ritual practices have been likewise classified as either traditional/indigenous or Christian/modern. Hasu (1999: 42) notes that in contemporary popular consciousness "[t]o be Chagga is to be Christian". This eventuality has involved the processes of accommodation and syncretism. Traditions, customs and rituals were re-evaluated as positive/suitable or negative/unsuitable for practice. For example, beer celebrations and sexual licentiousness were considered destructive to the Christian ideal of sober matrimony and thus strongly discouraged (1999: 216).

The indigenous conception of time was impacted upon through missionary and convert activity. The "rhythm of life" was Christianised (Pels 1999: 155). A distinct weekly and annual rhythm was marked upon the temporal scheme. The ceremonies of the Ecclesiastical year and Sabbath placed restrictions and prohibitions upon habitual patterns of behaviour. Whenever the mission had the influence it coordinated ceremonies and restricted business and commerce (Althaus 1992: 75). Chiefs soon refrained from performing their legal procedures during sanctified periods and markets were arranged for alternative days (Hasu 1999: 152). Alongside the Christianisation of time being Christian involved other behavioural prerequisites of conformity including reduced alcohol consumption, loitering and other negatively categorised sinful behaviour (Comaroff and Comaroff 1997: 64; Hasu 1999: 152). Moreover the body became the site of religious indigenisation. The bodily surface, ornamentation and clothing are symbolic media that perform the dimensions of the cultural and the social. The treatment of the "domesticated physique was an everyday sacrament" (Comaroff and Comaroff 1997: 220) that conduced a form of bondage. This theme has been explored in relation to contact and history in various African contexts (see Comaroff 1985; Comaroff and Comaroff 1992; Hendrickson 1996). The missions conducted much influence through establishing economies of consumption and commodification.

> "They…[colonies]…relied heavily on the circulation of stylized objects, on disseminating desire, on manufacturing demand, on conjuring up dependencies…[T]he banality of imperialism, the mundanities that made it so ineffably real, ought not to be underestimated" (Comaroff and Comaroff 1997: 219-20)

It has been cogently highlighted that in Kilimanjaro the missions used clothing as an arena of contestation whereby definition and categorisation were enacted (Hasu 1999: 196). Religious adherence and conformity to the colonial project was emblematically performed with the acquisition of acceptable bodily forms.

"Besides being clothed, clean, neat and orderly, this body incorporated appropriate habits, comportments and gestures indicative of a disciplined Christian whose interior morality was consistent with the outer body" (1999: 196-7)

Historically this was developed through missionisation in particular the efforts of the church to provision modern clothing to converts (Ntiro 1953: 29). Associated with this was the fact that rare clothing or ornamentation informed status. The origins of these performative costumes of rank and hierarchy stemmed from the commercialisation of the Swahili caravan trade during the later nineteenth century (Koponen 1988) and were exacerbated by the colonial discourses. In sum the morally sound bodies of the Christianised contrasted abruptly with the unsuitable African body that was naked, dirty, highly adorned with paint and traditional ornamentation. Even in the contemporary period physical hygiene is deemed to be related to morality. It has been noted that during this time the body was also the site of contestation through parody. The indigenous body appropriated foreign tastes and items according to pre-existing perceptions (Comaroff and Comaroff 1997: 236; Hasu 1999: 200; Martin 1994: 405). The desire for commodities was condemned by the missionaries in terms of the Protestant ethic and moral code. Material gain was conceived as being contradictory to idealism and redemption (Hasu 1999: 202, 216). In line with the Germanic *Kultur* theory adopted by the Protestant discourses money was viewed as precipitating negative social transformation. The missionaries perceived that money promoted individualistic tendencies and fragmented sociality through migration, bodily desire and selfishness (see Simmel 1990). Thus traits that might be erroneously identified as indigenous communalism were actually fostered by the missionising project. Indeed sometimes it is difficult to reconcile the subtle differences inherent in what is locally deemed positive traditional sociality and those that are negative characterised as conspicuous consumption and commodification (Hasu 1999: 371).

A marked exception to materialist possession was clothing for not only are clothes necessary they are highly significant means of communicating adherence and spiritual conversion. The missionary Gutmann reported "clothing was the language of the soul of the people" (Gutmann 1924: 181, in Hasu 1999: 204). It is interesting that traditional clothes are thought appropriate for funerals and European ones for other ceremonies such as weddings (Hasu 1999: 402). Obviously clothes, gestures and temporal orders coincided as expressions of culture at all times but particularly on Sundays. The missionary position, no pun intended, emphasised the relation between property and marriage. During the colonial period the performance of the Christian marriage ritual was only deemed appropriate if the Wachagga male had realised cultural expectations of house ownership.

Within this house the wife can be placed (1999: 367). Thus the religious potency of the landscape is stressed with the understanding that property and fixity relates to a Christianised mode of being. Moreover according to the missionising project sexual habits were culturally understood with metaphors of space. Women and the land have procreative potential that should be accessed through invocations and rituals. Thus abortions are perceived as the greatest form of sin – such practices are believed to destroy sociality and subsequent reproductivity. Abortions and contraceptives are believed to encourage women to roam around (Hasu 1999: 394). Roaming in this context is immoral and refers to increased sexual liaisons and contrasts with the Christian ethic of fixed monogamous matrimony. Women should be placed within a house. Indeed the moral female body is sometimes described in terms of a house with the sexual organs being the hearth (Emanatian 1996: 208). The immoral female body is mobile, loose, fast and commercial and is often described in terms of modern transportation (Weiss 1993: 22).

CONTEMPORARY RELIGIOUS IDENTITIES AND MEANINGS: EVIDENCE OF INTEGRATIVE MISSIONISATION

"People believe what surrounds them. When I was born I found the mountain and I found the church. I believe in the mountain as I believe in the ancestors and the Holy Trinity. This is a very obvious thing" (MC/Fo19/July 2004).

The traditional deity *Ruwa* is thought to abstain from direct intervention in humanly conduct unless extreme circumstances exist (Dundas 1968: 107). Thus the patriclans consider that it is wiser to conciliate the second-order supernatural entities that habitually interfere with daily existence (1968: 123). The processes of Christianisation and missionisation have had less effect than might be supposed in eroding away the importance of the second-order spiritual entities. The term *Mungu* is often used in place of *Ruwu* but in the popular consciousness and common valence the terms are surprisingly synonymous.

The contemporary Wachagga religious cosmology is composed of multiple historically appropriated elements that incorporate Christianity, witchcraft, ancestors and spirits (Hasu 1999: 519). The Wachagga identity manoeuvres between the political strategies of Christianity, modernity and traditionality. One does not only define oneself as Christian according to contrasts with the traditional or modern but more so against other religious and ethnic/tribal groups (1999: 406). Lifecycle rituals tend to be performed throughout *Uchagga* during the Christmas period (1999: 31). During this period the migrant workforce returns home to perform both Christian and traditional ceremonies.

This demonstrates the potency of the traditional rituals – cultural conformity is being facilitated in innovative contexts. Another factor that has been subsumed within indigenous matrices of cultural understandings is the AIDS pandemic. This event is understood in reference to biblical metaphors and evangelical teachings. The Christianised body of a Mchagga must be physically and sexually placed within the ancestral landscape to avoid judgement. Because there is no bio-medical cure it is logically perceived that the only protection against AIDS infection is to live the morally upright life espoused by the bible (Hasu 1999: 404). Essentially those that roam sexually harvest death.

> "Do not be deceived: God cannot be mocked. A man reaps what he sows. The one who sows to please his sinful nature, from that nature will reap destruction; the one who sows to please the Spirit, from the Spirit will reap eternal life" (Poster, quoted in Hasu 1999: 404)

In the indigenous discourses to transgress against one's own body is sinful. The body is the temple of the Holy Spirit and therefore to sin against it is to sin against God (1999: 405). One's personhood is possessed by the community, the divine and the ancestors. This partability of the body in life is continually reinforced and rearticulated in mortuary rituals.

Some level of synthesis between traditional beliefs and missionary teachings is to be expected. The missionary institutions developed intricate and sophisticated ways to facilitate and encourage the conversion of indigenous people throughout the world. A particularly successful method was to harness the potency of traditional beliefs by highlighting areas of commonality between traditional and new religions. Subconscious and ritual acts were subsumed within the new order of things. Such practice can be clearly highlighted with examples from the various areas of the study. Within the performative missionary instruments used to 'enculture' the Wachagga numerous traditions are engraved. Thus traditional beliefs are still apparent. The study frequently highlighted people who held traditional beliefs. Most would not declare their involvement in such practices but would with complete confidence state that such beliefs were commonplace.

> "If you believe in the mountain you come to no harm. Even now they do believe. Believing may mean going regularly and practicing other rituals like slaughtering goats and cows" (MR/As3/July 2004)

So universal was this supposed adherence that suspicions were raised that perhaps the locals were fabricating answers to interview questions, attempting to provide the information they thought the researchers wanted to hear or trying to implicate neighbours in acts of paganism. It should be mentioned that the indigenous villagers interviewed on all transects referred to the practitioners of traditional religions as pagans. Locally this term is synonymous with unholy, ungodly and impious. Any practice not compatible with the world religions, in particular Christianity, integral to the area were conceptualised as a form of paganism. Suspicions concerning such implications and fabrications were proved incorrect, however, when multiple informants admitted their involvement in some traditional practices and the preservation of certain beliefs. Moreover interviews and conversations with local religious leaders confirmed the situation although their representation of the locality did not directly correspond to that which availed itself during the study. They thought the estimations based on participatory fieldwork, into the prevalence of traditional belief and ritual, were exaggerated. Admittedly they are likely to have better comprehension of the local situation, especially considering their vested interest in such matters, but one is unlikely to willingly reveal one's involvement in practices frowned upon by the church to church authorities. Likewise the church elders and fervent types within the village are well known within these tight-knit communities. Traditional practices within the Lutheran transect areas – Marangu and Machame – were very much covert activities. Thus one would not advertise one's association with these activities. It is also worth noting that the churches were unlikely to fully acknowledge the situation if they view such beliefs as deviant. This could have performative effects. Indeed traditional belief/practice on all transects were negatively categorised by the various churches and their representatives. This has been effective in marginalising and peripheralising overt practitioners of traditional religion.

On the Rombo transect what was locally termed 'paganism' was more overt. These practitioners of traditional religions were also far more comfortable to converse about their involvement in such things. Moreover those inhabitants who were 'adherents' to the world religions that were uncomfortable talking about their implication in such activities from the outset became more confident in revealing certain things after a level of trust had developed. It transpired that adherents to all the world religions present in the area – in particular Christianity and Islam – were practicing elements of the traditional religions with varying degrees of covertness. A common belief in the area was that if one was caught practising traditional religions by a clergyman or church elder then a curse would be put on them and their family and they would die. This extreme conviction can be explained in terms of disassociation and the deprivation of local religious facilities such as schools and hospitals. Yet it also demonstrates the belief that the religious leader would actually curse them.

"The old ways were better than the new ways. To be pagan was better than being a Christian. We used to slaughter goats and cows and then it would rain and the land produced much food. Now with Christianity we get no rain or crops. So I teach my grandchildren the old ways" (MH/Mk4/August 2004)

During various informal conversations and formal interviews with various individuals the question was frequently asked 'do you not have traditional religion in England?' It soon became apparent this question was being asked in response to my inquisition regarding the traditional religion because the interviewees were failing to understand what they interpreted as a lack of understanding. There frame of reference was perceived as being not only obvious but simply the way things are. Religions are replaced by new versions but what works locally or what makes the world coherent retains believers. One pagan, the term is deemed appropriate in this context because that is how he described himself, claimed that he would pray to *Ruwa* through the ancestors and in the morning he would face the mountain and pray. He also revealed that daily pressure to convert is applied onto him and his family by the leaders of the Roman Catholic, Lutheran and Pentecostal churches in the area. Indeed he had to pay to have four of his "eleven children baptised so that they can go to the school" (MH/Mh24/August 2004).

The early missionaries translated German hymns into Kichagga (Hasu 1999: 463). This probably related to the indigenous fondness of song and its centrality in both ritual and everyday life. The translation of hymns was given a high priority for similar reasons in the Anglo-Saxon missionary schools (Comaroff and Comaroff 1991: 241). The popularity of such works increased when environmentally specific allusions were included within the hymns making them pertinent to the indigenous existence. This is further evidence for this adaptionist and syncretistic agenda. For instance, the following two extracts are typical of hymns from an indigenous hymn book dating to the early twentieth century. These hymns are still popular on the slopes today. Whilst conducting the interviews and covering such issues numerous participants would spontaneously sing these and other similar examples. They form part of the religious expression of the locale. The following Kimashami text is taken from a Lutheran publication (Evangelical Lutheran Church in Tanzania 1908) and the translation was accomplished with the assistance of multiple local informants and interpreters.

202. Schönster Herr Jesu (The Goodness of Jesus)

Kyaamwi kyashia,
mmbo kirooye uuwe,
kikakooya mafishi.
Yesu nsha torya

Yesu m mwaa torya,
akee mwaruta mirima.

The mountain is the holder of the light,
It is beautiful, upright and high,
It reaches the clouds.
But Jesus is more beautiful above all beauty
And Jesus is so bright above all brightness
And you make the heart and soul happy.

Mwiiri washia
na mwi na nyingiri
fivaa fikaluseta.
Yesu afiirya
avaa kilanya
ya vasu vaa va Irava.

The moon, the sun, the stars
The brightness, the light and the atmosphere
They make us happy.
But Jesus is above them all
He shines the greatest
More than the angels of God.

265. Meinen Jesum Laß Ich Nicht (My Jesus Leave Me Not)

Nkundye iva Yesu-fo,
Ankuvika ando akwa.
Kuti shikumwosire
sha urovirovi tapu.
We ni sa ivaiyaa muu.
Nkundye ira Yesu-fo.

I do not want to leave Jesus,
For He has acted on my behalf
I must embrace Him
Like urovirovi (clinging grass).
You hold so firmly as the light of life
I do not want to leave you my Jesus.

Essentially these hymns represent the manner in which the local environment and the memoryscape were being recontextualised and adapted within the new cultural and religious project. This incorporation makes the new religion simply a novel rearticulation of traditional awareness and of indigenous knowledge systems. Hence the new religion and its practices contained a high degree of familiarity for those undergoing conversion i.e. it was more likely to make intuitive sense. God and Jesus are not alternatives to the sun, the moon, Kibo or other elements of the landscape but relate to their potency. The mountain (see **figure 2**) is not the seat of the divine but demonstrates the power of the sacred. The hymns reiterate that the Christianised *Ruwa* surpasses the traditional religious foci in every tangible way but do so by juxtaposing them and thus unavoidably relating them. Moreover the expression of symbolic concepts such as

23

entanglement, unity and adherence required innovative strategies for ensuring full comprehension. In hymn 265, for instance, a simile is utilised to better facilitate understanding – thus one is instructed to cling to the divine like *urovirovi*. This form of tangle-weed is very common to the forest belt of Kilimanjaro and all local people are very familiar with it (**figure 3**).

Perhaps an even more indicative case concerns the axial alignments of church buildings. In the Machame and Maharo areas pilot studies were conducted upon the local churches focusing on their axial relation to the mountain which, as previously mentioned, was the seat of the traditional divine (see **figure 4**). It was hypothesised that such relation to the mountain would belie the missionaries' methods of eliciting/finessing conversion through integrative accommodation. Observational survey established from the outset that churches tended to be sited on ground of high relief. This was obviously a strategic decision by the missionaries to dominate the neighbouring landscape. Some informants thought it might have some relation to the height of the mountain i.e. the churches reach up toward it (e.g. MC/Fo1/July 2004). To be high is to be divine. It is also interesting to note that the present and former Mkuu Parish Church buildings were constructed upon the old mission. Moreover the mission itself was originally built upon a traditional sacred site (MH/Mk18/August 2004). This was undoubtedly a deliberate action on the part of the early missionaries to rearticulate the inherent power of the place into a coherence they would communicate through biblical teachings. This deliberate reuse rearticulated the inherent power of the place into a coherence that was communicated through biblical teachings. The potency of certain points of the landscape was harnessed by the missionaries. The historic dimension to these sites embellished the memoryscape.

This survey included all the churches in the wards of Machame and Maharo that were bisected by the interview transects (**for examples see figures 5.1, 5.2, 5.3 and 5.4**). Despite the small number of churches actually surveyed (n=13) the study raises some potentially interesting questions about the motivations behind local architectural design and setting. Clearly church Altars are aligned with the mountain. Whether intentional or not, and it is highly likely that such building projects were thoroughly applied to the local context to maximise effect, symbolic reasoning for the alignment cannot be escaped. No Mchagga would be able to unconsciously or consciously escape its permeation.

Figure 2. Photograph of urovirovi (tangle-weed) on concrete for contrast

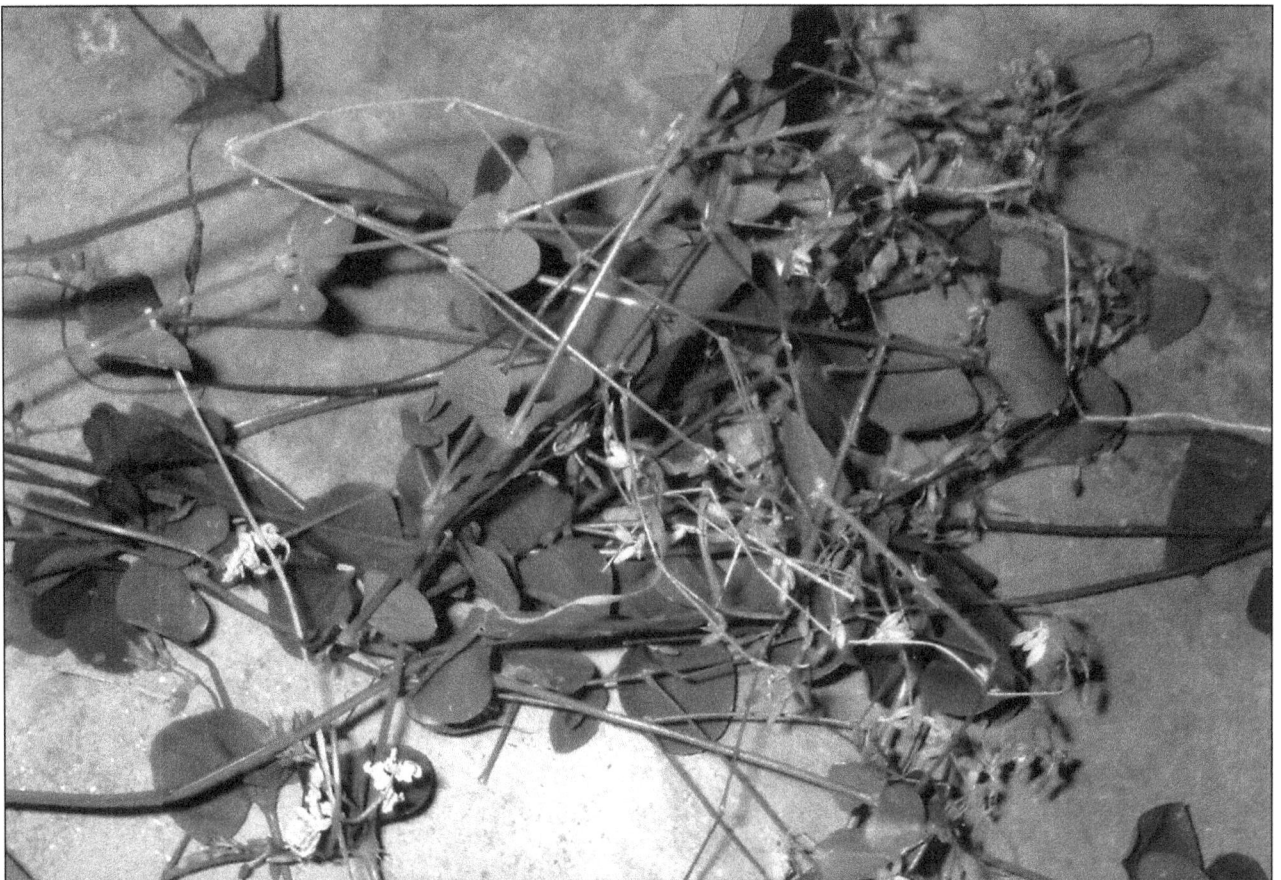

Figure 3. Photograph of Kibo from position in the Maharo study area

The consciously orchestrated positionality of the churches and resultant Altar-Kibo axial correspondence related, and indeed relates, the traditional *Ruwa* with the new missionary deity. As noted these architectural decisions had further useful performative implications: rituals were choreographed in a manner that maintained some intuitive action and thought based on habitual rehearsal. Moreover the newly introduced practices conformed to a great extent to the local historical, cultural and ritual *habitus*. An illustrative example of this concerns praying at the Altar. Traditionally much ritual action incorporated the mountain in some fashion – as a direction, as a point of reference, as a religious symbol and so forth. A famous and widely recorded example notes that during the precolonial period a Mchagga would, upon awakening in the morning, spit saliva toward the mountain (Mosha 2000: 77). With conversion came the required performance of new rituals such as prayer, partaking in communion, hymn-singing, genuflection, and reciprocated engagement with the rites of preachers, missionaries and evangelists. These rituals incorporated the divine within their practice and by having the Altar aligned with Kibo adjustment would have been subtle. In many ways one was relating to the numinous in a comparable manner. Previously godly engagement involved the mountain and subsequently it

involved the Altar. In simple terms one could still pray to *Ruwa* through the mountain whilst adopting the ritual practices of the new religion.

Much further research needs to be done regarding the phenomenon of Altar-Kibo axial relation in the Kilimanjaro region – how widespread was/is the practice, can different policies be discerned between the competing religious missionary factions and their building projects, what if any are the engineering implications of such relation. Why is there a trend for a subtle 5-10° misalignment in the Machame area? It is suspected that this owes more to engineering and construction consideration rather than strategy but such speculation requires substantiation.

Another arena of indigenised Christianity concerns witchcraft. Witchcraft and Christianity are mutually intelligible in their culturally dominant concepts of desire and jealousy. The historic and contemporary witchcraft discourses offer an idiom of choice for the comprehension and orchestration of change and development (Fisiy and Geschiere 1996: 194; Hasu 1999: 449). Publicly all the Kilimanjaro churches condemn witchcraft as inherently sinful. In fact considerable effort from churches and even local government has been

25

Figure 4. Table showing the orientation of local churches in study areas

English name	Kiswahili name	Kimashami name	Village	Religious Denomination	Alignment of Altar to Kibo (degrees)		
					Current	Previous	Closest
Believe - In	Nkwarungo Kusirye-U	Muoamini	Foo	Lutheran	120°	5°	5°
I Am With You	Niko Pamosa Nanyi	Shikee-Neeni	Foo	Lutheran	10°	N/A	10°
Jesus Our Saviour	Yesu Kristo Nimwokozi	Yesu Nyi Nkira	Foo	Lutheran	315°	N/A	45°
Church of the Heart	Kamisa La Kiroho	Nyikamisa Lya Mrima	Foo	Pentecostal	15°	N/A	15°
N/A	N/A	Nkwasanu	Nronga	Lutheran	15°	N/A	15°
Continue to Praise	Mtukuze	Arafumin	Nronga	Lutheran	10°	N/A	10°
N/A	TEWO	Kanisala	Nronga	Pentecostal	355°	N/A	5°
He Is the Truth	Ni-Wa Kweli	Mbwa Dede	Wari	Lutheran	355°	N/A	5°
Door to Heaven	Betheli	N/A	Wari	Lutheran	180°	N/A	180°
The First and the Last	Alfa Na Omega	Wa Mwanzo Na Mwisho	Uduru	Lutheran	0°	N/A	0°
Mkuu Parish Church	N/A	N/A	Maharo	Roman Catholic	0°	N/A	0°
Old Mkuu Parish Church	N/A	N/A	Maharo	Roman Catholic	0°	N/A	0°
Kirokomu Parish Church	N/A	N/A	Makiidi	Roman Catholic	180°	N/A	180°

invested to discourage the work of witchdoctors and spirit exorcists (Hasu 1999: 440). Conversely there is plenty of evidence that suggests local authorities and clergy have facilitated the presence of certain traditional medical practitioners (1999: 443). It is interesting to note that the monetisation that accompanied the colonial project actually exacerbated the influence of witchcraft as substances and instruments could be accessed by all (1999: 523). Moreover the contemporary cultural phenomenon of AIDS has further increased the dependence on witchcraft and other occult activities to the extent that some are recognising it as a postcolonial phenomenon (e.g. Fisiy and Geschiere 1996).

Witchcraft substances relate to the body through the environment. In the past actual body parts from graves were frequently used and in many cases were consumed. This necrophagic practice is considered very potent in the rituals of spirit possession and exorcism. Other substances included plants, animals and other things that are connected with a particular person through physical or symbolic associations. One can argue that traditional Wachagga witchcraft or *uchawi* (see Marealle 1947) has been eradicated. The traditional beliefs involved an inherited capacity passed through the female lineage. The witchcraft notions currently enjoying currency involve the purchasing of spirits or *majini*. The *majini* is not bought locally but from the coastal Muslim areas (Hasu 1999: 411). These spirits are purchased and are known to be foreign. According to local knowledge the *majini* are cultivated in forest areas near the coast (1999: 436). The Wachagga believe in two contrasting and unrelated forms of spirits. On the one hand the spirits that embody places which are understood to correspond to the divine and on the other the *majini* that are associated with satanic forces and thus inherently bad. This is of course another example of the indigenisation of Christianity. It is also important to note that AIDS deaths are often explained in terms of sorcery, spirit possession and witchcraft (1999: 415). This is in part due to scientific illiteracy, social embarrassment, and of course strong beliefs in the powers of witches. The main types of *majini* recognised by the Wachagga are those that feed on humans by sucking out their blood and those that intrude into the body and interfere with it (1999: 428). During the advanced stages of HIV infection body weight drastically decreases and it is locally interpreted that blood sucking is responsible (1999: 430).

Figure 5.1. Photograph of Nyikamisa Lya Mrima Pentecostal Church, Machame

Figure 5.2. Photograph of Muoamini Lutheran Church, Machame

Figure 5.3. Photograph of Mkuu Roman Catholic Church, Maharo

Figure 5.4. Photograph of Old Mkuu Roman Catholic Church, Maharo

CONTEMPORARY AND HISTORICAL FUNERARY RITES

It was noted at the time of the British colonial administration that adherence to traditional beliefs and practices was not uniform and that there were many individuals even elders sceptical of their practices (Dundas 1968: 179). Obviously processes of missionisation had influenced the indigenous people but certain beliefs remained stronger albeit in rearticulated and less overt manifestations. Moreover, confidence in the traditional religious practices was noted to increase during periods of untoward and stressful events (1968: 179). There are many recorded incidences of increased confidence in indigenous beliefs at the expense of the missionary teachings. For instance after a period of severe drought which was exacerbated when local cattle began giving birth to stillborn in Masama it was reported that local farmers, believing the ancestors were responsible, asserted that the converted would soon tire of their new beliefs (Jessen 1909: 61, in Hasu 1999: 190). Moreover others supposed that modern technologies such as the telegraph line were the origin of such misfortunes because they interfered with rainfall (Weishaupt 1912: 431). The *kihamba* or banana grove (more below) is sacred land and this religious dimension is surmised well in the words of Dundas (1968: 194) when he explains that "the initiated tread there with reverence, knowing that the grove is first and foremost the family graveyard". Western-style graveyards were established in numerous areas from about 1900 and were utilised by Europeans and indigenous converts alike (Gutmann and Jessen 1905: 490; Hasu 1999: 151). Nonetheless the banana grove remains the normative site of burial. Nowadays the church blesses land to be used as grave plots (Hasu 1999: 460).

Historically after death the corpse was stripped of clothing and ornamentation and the legs and head were bound together in preparation for burial (Dundas 1968: 181). Under the early missionary influence the custom of wrapping the corpse in animal skins became commonplace and in the contemporary period a shroud is utilised. The shroud should originate from the home of the deceased just as formerly the animal skin should have been derived from a head of household cattle (Hasu 1999: 459). In the past the primary burial involved an undisturbed interment within the familial hut. The grave was dug with a branch of *msale* (*Dracaena steudneri*). It was believed that the spirit of the deceased was confined whilst within the grave. After the burial a vigil was kept around the grave for numerous days until the ceremonial hair shaving was performed. Hair-shaving involved close male and female relatives and if married wives that were unlikely to remarry (Dundas 1968: 184). During the fieldwork evidence of this practice was still discernible (MH/Mh12/August 2004). Cursing those responsible for a death has also been a common part of the funerary rituals since the colonial era (see Dundas 1968: 187) and one suspects long before. After the primary burial there

was a period when further sacrifices were usually made. This involved the memorialisation of the deceased (Dundas 1968: 188).

In the Wachagga cosmological order death is traditionally associated with fertility. This was articulated in the traditional treatment of the dead bodies of childless and uninitiated adults that were discarded in the forest belt so as to not disturb the cosmic order of the ancestral land. Moreover the bodies of babies and the uncircumcised were discarded unburied within the *kihamba* (Moore 1977: 47). The treatment of the circumcised bodies of those that had borne or sired children was much different. It was a kindred duty to ensure that the dead were remembered in this world and that they had everything required for their journey to the ancestral world (1977: 69). The female dead were buried in their huts and the male in the hut of his first wife. The symbolic importance of *msale* is again clearly emphasised in the ritual activities that utilised it. The graves of both men and women had to be dug with the branches of the tree as instruments. Prior to interment the male body was anointed with red ochre, butter and *msale* leaves. The body was then covered in banana leaves and/or the hide of a recently slaughtered ox. The body was ritually positioned in the grave in that it was either laid or seated on its right (masculine) side facing the peak of the mountain (1977: 69). It is reported that several persons maintained a vigil on the grave for the next four days eating the meat of the slaughtered ox or another slaughtered animal. For these days those present were required to talk about the dead person – ritualised memory production – while they ate thus ensuring that the deceased was remembered.

It was also the responsibility of the vigil keepers to conduct cursing ceremonies. In particular on completion of the remembrance period they had to orientate themselves toward *Kibo* and curse four times anybody who may have been responsible for the death (1977: 69). The spirit confined within the primary burial site required release from its confinement after the disintegration of the body. This precipitated the secondary burial. This occurred between one and two years after the primary entombment. Access to the grave was from the outside of the hut which involved tunnelling using an improvised tool fashioned from *msale* (Dundas 1968: 190-2). The skull bound to the humerus was buried or planted at the patriclan *mbuoni* facing in the direction of Kibo peak. In Uru skulls were frequently buried in earthenware pots whilst the norm elsewhere was burial without covering (1968: 192). Stones were often placed above the skull burial and were "carefully left undisturbed" (1968: 193). It is interesting to note that in the contemporary era although the tradition of secondary burial has been eradicated stone relocation is still very much the norm. According to some 'double burials' have not been practiced for decades (Hasu 1999: 457). Interviews suggested that elements of the transformative rites are still conducted and the rearticulated bones of ancestral bodies are still indigenously considered potent ritual objects (MR/Ly10/August 2004).

Precolonial mortuary rites required that the *msale* plant was cultivated throughout the skull grove (Moore 1977: 49). The bones were haphazardly discarded throughout the grove with the exception of the skull and the humerus. These were positioned with care. The skull was bound to the humerus with *msale* and then positioned in the ground facing *Kibo* amongst the skulls of the other ancestors (1977: 70). It is also interesting to note that missionary activity was thought to have abated local fear of witchcraft (Bailey 1968: 167). Witchdoctors were traditionally thought to be the "voice of Ruwa" (Marealle 1965: 60). This contradicts more recent research that notes the pervasiveness of traditional rituals, myths and powers of bewitchment. Indeed witchdoctors commonly use the *msale* plant in divination and in drawing-out bewitching spirits (Mosha 2000: 241). There are many commonalities between the traditional funerary practice described here and that which is conducted presently.

As with other traditional rituals those that neglect performances, which can be commonplace if one's 'modern' faith is particularly strong, are thought to precipitate misfortune upon all (Hasu 1999: 450). In fact these assertions are often articulated with similar charges of modern misguidance. For example, those that are identified as being successful in the modern ways of the *kizungu* i.e. have money, education and professional employment will die of AIDS because of increased access to sexual partners (1999: 451). The abandonment of traditional rituals has been linked to the hopelessness brought about with the AIDS pandemic. In this sense to be modern is a deadly and perhaps accounts for some resistance to change. Death is an everyday occurrence in *Uchagga* and thus mortuary and associated rituals have "become the rituals of modernity" (1999: 527).

The treatment of a lifeless body by the living corresponds to the expectations of the living and the soul of the deceased (Huntington and Metcalfe 1979). Relatives expect a displaced Mchagga to be buried at home. In many ways to be Wachagga is to be born, to live, and especially to be buried in the *kihamba*. Wachagga mortuary practices are transformative rituals of placement and orientation. The dead are transformed into the spirit world. In death one must be connected to the spiritual plane of the ancestral lands. Bodies are unmade through the process of burial. After departing the house for the final time, which is believed to be a feminine and social arena, the unmaking is enacted through sacrificial action. Moreover due to complex webs of relations the 'unmaking process' involves larger gatherings when the deceased was of social and reproductive maturity (Hasu 1999: 466). It is also interesting to note that the umbilical cords of Wachagga infants are still buried in the *kihamba* even if they are born far away from the ancestral home (1999: 450). The umbilicus is dried for a few days and

then buried with manure under a specific banana tree. The generations are ritually integrated with subsequent consumption as the ensuing fruits of the banana tree are to be consumed exclusively by the grandmother of the infant (1999: 473). This act begins the process of making the individual. Thus birth and death place the individual within the ancestral landscape. The *kihamba* is the beginning and the end of the bodily person (1999: 472). This process of bodily growth is staged upon the land – the soft black umbilicus is transformed into the hard white skull. The umbilicus transforms into the buried corpse. The past, present and future are physically manifest in the patriclan's *kihamba* (McCall 1995: 259). The burial rituals are acts of placement that both make and unmake the person. The generations of the *kihamba* are consubstantial in that one generation is the source of vitality, substance and life for the next (Hasu 1999: 477-8). Some have suggested that the modern lifestyles, linked to the AIDS issue, fail to reproduce the notions of personhood that integrate belonging with personhood (Setel 1995a, 1995b, 1996, 1999) but these misrepresent the situation. Such personhood is continually reproduced within matrices of local understandings and perceptions.

Wachagga mortuary rituals, despite encouragement from the Lutheran church to shorten the timeframe, are performed over several days (Hasu 1999: 457). It is customary for relatives to abstain from any domestic or remunerated work from the time of death until the mortuary rituals are complete (Hasu 1999: 458). The grave is dug and the burial is conducted on the first day and a sacrifice is performed on the second day. It is thought that the grave should not be dug and left open overnight otherwise it will consume other persons. Burial plots are spatially organised to denote social positioning e.g. the graves of those who have not fulfilled their marital and reproductive potential are peripherally located (1999: 460). The corpse must visit the home before being buried. This "act of returning the deceased into the house parallels with the former secondary burial" (1999: 461). There are other traditional elements referenced by the Christianised mortuary practice. Vaseline and/or butter are applied to the corpse so that they might be received well at their destination – the metaphoric journey to the afterlife. In the past the body would have been smeared in animal fat to denote an apologetic offering to the ancestors (1999: 462). Moreover the corpse would have been buried with either a spear if male or a cooking instrument if female. If a woman was pregnant or breastfeeding she would also have been buried with a banana flower. The church strongly discourages these practices by threatening excommunication and monitoring the sealing of the coffin to avoid inclusion of foreign objects (1999: 462-3). Also whilst the body is being prepared for burial spray deodorant is liberally released into the air. This probably relates to the former practice of skeletal exhumation in preparation for secondary burial when the odour of decay was forceful and necessitated access from the exterior of the hut (1999: 463). *Msale* is planted on the four corners of the grave which culturally resonates and amplifies the former practice at the *mbuoni*.

It has been reported that some Wachagga, usually those that have been excommunicated from the church, are buried in the *kienyeji* or traditional manner. These are not burials in the style of historic ceremonies but rather irreligious episodes. The *kienyeji* burial is not thought to facilitate incorporation with the ancestors, which requires a Christianised funeral, and is thus considered horrifying. Christianity has been indigenised to the extent that the person cannot be unmade without its content (Hasu 1999: 484-6). In order to become an ancestor one must be buried according to the Christian practice and two sacrifices must be conducted (1999: 518). Therefore the emotional affinity between communities and ancestral territories is an important motivational force within social groups. The dwelling of the living must be made in the dwelling of the deceased. With death an individual must be returned to the landscape and to the ancestors (Hasu 1999: 452; Lan 1985: 20). It is commonly believed that deceased people never leave the area where they died unless certain rituals are performed to facilitate the movement of the spirits. Therefore there is a permanency in an individual's and a patriclan's attachment to land. These spirits always have a local dimension to them and so are potential media between the physical and spiritual planes of existence. Thus burials and rearticulation of bones in the *mbuoni* keep the spirit close to the domestic area of the related living so as to better assist and inform them. The bones are no longer rearticulated after decomposition in the grave instead they remain in the grave. Nevertheless most graves are on the domestic farmstead and so the spirits remain in close proximity. The dead are very much believed to influence the living. They communicate. Furthermore the dead converse (maybe through the divine or for the divine) through the environment.

Wachagga funerary ritual activity is complex and demonstrates a remarkable level of syncretism between the traditional and missionary religions. The following description, tempered with informant collaboration and substantiation, concerns a funeral and associated rituals as practiced by a Wachagga Roman Catholic family in the Rombo district. Directly following death there is a three to four day preparation period for arrangements and grief. Following that the burial ceremony takes place. The *kipata* or grave plot is usually on the family land. In this sense the family and patriclan life-cycle is performed upon a distinct portion of the landscape and this sediments identity. The funeral service, conducted by a clergyman, is essentially the same as any other Roman Catholic service (MH/Mh41/September 2004). After the funeral there is a 'sorrow day' before the commencement of the traditional slaughtering. These traditional practices are referred to by the church as 'customs' rather than rituals or religious activity. Presumably this is the best way to disassociate themselves from the proceedings.

After the funeral men who were familiar with the deceased stay around the house for three days whilst women stay for four days. Women congregate inside the house whilst men remain outside. The women traditionally prepare *kiburu* (a soupy porridge of beans, banana and water) which is a ceremonial drink usually made and consumed at funerals. The preparation of this drink involves a cursing ritual. The women come out of the house and sort out beans and *mbege* (finger millet) upon a hide on the ground. Once this task is completed all the women take hold of the hide and rotate/circulate it whilst chanting in local dialect "if someone has caused this death let him die". Then the hide is reversed and hit against the ground. This ritual is completed with water being used to wash the area outside of the house just involved in the preparation of *kiburu*. During this time the males will be cutting banana leaves for use in the preparation of *wari* or local brew (banana beer). The slaughter is then ready to be performed. A sacrificial animal is brought out of the house to the *kiungu* or sacrificial site (more below). The *kiungu* has been prepared for the slaughter with banana leaves laid out. The men assemble to participate in the episode. Initially the goat is semi-suffocated and then its throat is cut. The gushing arterial blood is caught in a bowl and one man will stir this with his hand to prevent it clotting. The blood is later added to the *kisusio* or funerary soup. The making of *kisusio* is a male event. The animal carcass is then skinned and butchered by a knowledgeable man usually one who was related to or a friend of the deceased. Every edible part of the goat is utilised. After completion the meat sections are laid out on fresh banana leaves (see **figure 6**). Small pieces of each meat section are added to the *kisusio* at this point including scalp, testicles, colon, blood, skin, rump, chest and limbs. The head is taken inside of the house where for the next two days it will be deposited. After these two days it will be placed just outside the house and that is where it will remain. The food is then cooked by the women and divided according to customary law. The elders receive the chest and a leg, the related women receive the rump and a leg, female siblings receive the upper ribs, male siblings receive a leg and the non-related attendees receive a leg and any other remaining meat. Everybody present consumes some *kisusio* which has been cooked in the sacrificial area (**figure 7**). After the sacrifice has been consumed all the related women have their heads shaved and the *kiburu* is consumed. After the *kiburu* is taken the domestic areas can be swept for the first time since the death some five days or more earlier. One year from this day the *wari* that was started is consumed and a further slaughter is conducted. After this slaughter the deceased possessions are divided amongst friends and family. Traditionally two years after that the body would have been exhumed and the skull and skeletons would have been moved to appropriate *mbuoni*. The bones of the dead are sacred things and are shown the utmost respect. These days exhumation is rare, if practiced at all, and the usual custom is the relocation of a stone from the grave to the *mbuoni*. Nonetheless rumours of bone rearticulation abound. The *mbuoni* phase of the rituals sometimes also requires the slaughtering of further heads of cattle. The numbers involved seems to vary according to patriclan affiliation.

Figure 6. Photograph of sacrificial portions of meat placed upon freshly cut banana leaves

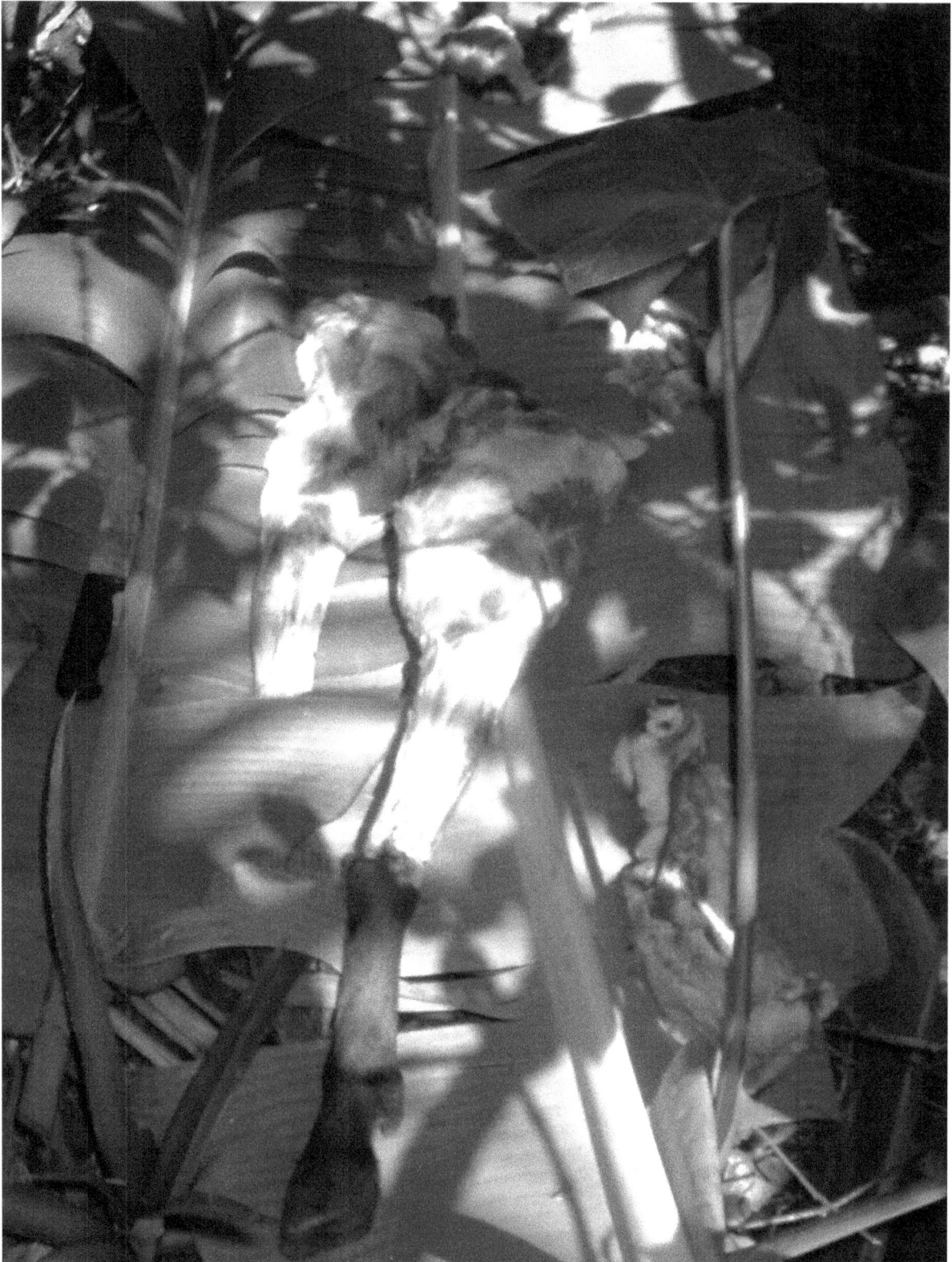

Figure 7. Photograph of funerary kisusio preparation

CHAPTER FIVE

PHYSICALITY OF KIBO

REVERENCE AND IRREDUCIBILITY

It has been noted that soils, seeds and water will commonly assume cosmological significance (Shipton 1994: 367). The environment is made intelligible through processes of reverence and vitality. The divine supernatural force that is *Ruwa* is incomprehensible and indescribable. This relates to one valiant attempt to conceptualise the numinous that notes the sacred is awe-inspiring and overpowering (see Otto 1950: 16-7). Indeed believing in the sacred is "like standing at the foot of a towering mountain" that precipitates the feeling and intuiting of awareness, humility, wonderment, reverence and gratitude (Mosha 2000: 9). The physicality and the presence of Kibo attach the Wachagga to the environment as a coordinating focus. Kibo dominates the landscape and everyone and everything sit in relation to it. Their being-somewhere corresponds to their positionality *vis-à-vis* the entity that is the mountain. In this sense the Wachagga are intrinsically linked to the mountain through emotional attachment that is manifested in bodily responses and supernatural experience. Rock as a material has a permanency beyond other natural and artificial types and therefore offers understandings of the past, present and future (Rainbird 2002; Taçon 1994)

It has been commented that it is impossible to represent the divine symbolically and consequently the Wachagga have no images of *Ruwa* (Mosha 2000: 8). This might seem to contradict the essence of conceptualisations and correlations between the mountain and the divine. However, such assertions can be countered admirably with the following remarks made by Paramount Chief Marealle II on opening an art exhibition

> "There are... so many facets of Kibo that a hundred artists could do the Old Man without finishing Him! I cannot think of a better subject for the brush that, and beginning there you have the rest of the world on your plate" (TNA/NA/L5/21)

The cultural sensitivity the Wachagga demonstrate toward the mountain is readily disclosed in the popular addresses of those holding public office. It is also interesting to note in the previous and following quotations the deference afforded through capitalisation and anthropomorphism.

> "Our buildings have Kibo behind us, and people may wonder why our Patron and

Protector should be so insulted. We are facing the future by looking at the expansion areas below, with our Protector behind us, and so Kibo is quite happy" (Chagga Council 1955: 35)

The Wachagga conceptualise Kibo in therianthropic and emotional ways. Kibo choreographs and fixates identities with its infinite significance, dimensions, indefinability, immeasurability, profoundness, irreducibility and numinality. Through its physicality the mountain is able to sustain the inhabitants of the slopes.

> "Of course I have plenty of reverence for the mountain – it supports and strengthens me as it supports all the Wachagga. When I return to find Kibo after being away my heart is warmed" (MS/M1/August 2004)

TRADITIONS, RITUAL MEMORY AND COSMOLOGY

The cosmological and ritual beliefs of the precolonial Wachagga reported in the literature suggest they were far more complex than first assumed by the early colonial and missionary migrants (see Moore 1977). Much of this traditional cosmology still impacts upon contemporary perception and is demonstrated in the cultural *habitus*. Landscape memory permeated every facet of the Wachagga cosmological world. The Eliadean dichotomy sacred/profane was not a feature of religious experience. Nothing in the world was without a sacred or supernatural element. Thus through their performance "all human activities had other-worldly significance" (Moore 1977: 46). Moreover nothing is inaccessible to spiritual intervention (Knudsen 2002: 21). Hence all ritual symbols were drawn from ordinary objects/categories. Moore (1977: 46-7) lists seven symbolic themes used in the ritualisation of the Wachagga worldview. These comprise the body, including its processes and products, basic foodstuffs, food processing activities, man-made objects and structures, the plants and animals of Kilimanjaro, natural phenomena, and natural dimensions. Some of these ritual categories continue to facilitate the employment of the mountain in ritualised activity. Furthermore "all things and processes were interpretable in terms of classifying categories that made supernatural sense out of the natural and cultural worlds" (1977: 49). Thus the mountain, implicated in the cultural world, was through these symbolic categories used to make sense of the supernatural.

The 'natural phenomenon' was one of the Wachagga symbolic classifications used in making sense of the world. Thus the mountain through its physicality becomes intelligible in terms of supernatural or enhanced natural agency. Furthermore the 'environmental dimension' was another cosmological indicator. Indeed the dichotomy higher/lower with higher being superior to lower is one such important dimension. The Wachagga demonstrate an occidented navigation or scheme of alignment – they take their orientation from the mountain and not the sun. This has become bound-up with the dimensional opposition up-mountain/down-mountain (Moore 1977: 49). It is well-documented that locally directions are given as 'up-slope' or 'down-slope' (Tagseth 2003: 14). Moreover it is customary upon a meeting that the individual coming from above – the direction of Kibo – gives greeting first because of associations of fortune and honour (Dundas 1968: 39). Social hierarchy is also similarly expressed. A superior individual should always be offered the higher and more honourable side of the road (1968: 39). Various other flora and fauna are incorporated into the cosmological and symbolic world associating the mountain with the supernatural world. For instance, to recognise the completion of the transaction/sale of *kihamba* land the vendor, in the presence of witnesses, will cut some leaves from a banana tree standing on the land in question and give them to the purchaser to take away (Johnston 1946: 7). Thus the symbolic leaves are part of a performed ritual of land transaction. The *kimanganu* ordeal is a further example of vegetation featuring in the mythological world. The ordeal was formerly a means used by the chief to settle disputes. The ordeal required certain subjects to consume an intoxicating and partially poisoning drink made from the dried leaves of certain plants (Moore 1970: 329). The concoction was believed to act like a truth drug and when the subject was questioned the chief could ascertain guilt or innocence. The ordeal was known to cause prolonged illness, swelling of the sexual organs and sterility in some of those it was administered to. Thus as Moore (1970: 329) notes "many an accused person confessed before being subjected to the ordeal". The elders warn the children to "be careful to tell the truth for the mountain will tell us if you are being false" (MR/Ly13/August 2004). Such panic-mongering amongst the dependent population may well relate to the *kimanganu* ordeal.

It has been noted that in pre-Christian times the *msale* plant had important "religious and social significance for the Chagga" (Bailey 1968: 166). It was regarded as symbolic of the afterlife and had many ritual uses. Bailey (1968: 166) notes that mission activity brought an end to the socio-religious significance of this plant. This is simply not the case. The plant is still used for demarcating boundaries, fodder for animals, and it is thought its inherent healing properties will cure certain ailments. The plant is also linked to forgiveness and pardoning.

"If I do you wrong but later come to you with *sale* leaves and break them… and rub them on your chest then you must forgive me and the wrong must be undone in the eyes of all. For if the bitterness remains then it is you that will die… [and] this is why the Wachagga are so forgiving a people" (MR/Mb8/ August 2004)

Msale is a pervasive signifier of death. This is because it plays an important role in death rituals. The Wachagga ritual response to death was multifaceted. This is because death is conceptualised in terms of a communicative vehicle bridging the supernatural and natural world (Moore 1977: 48). Moreover *msale* is encouraged to grow on specific sites associated with death (MS/M1/August 2004). It is also noted that traditional abortion practice is conducted with a branch from the *msale* tree being induced into the uterus (Hasu 1999: 479). This symbolic correspondence between death and *msale* derives from memories of traditional ritual and cultural practice. Moreover the plant *isale la shofu* (elephant *sale*) is believed to deliver, when consumed, a similar prophylactic effect to quinine for malaria. The belief may relate to the sacred qualities of *msale*. The medicinal dimensions of *isale la shofu* may be derived from this cultural tradition. For illnesses are frequently understood in non-scientific ways but rather as embodiments of bad spirits. Thus *isale la shofu* can bring death or life to the bewitched or poisoned body.

The fauna of Kilimanjaro is still involved in the ritualisation of the memoryscape. It has been noted that certain patriclans have totemic relations with certain species of animals or birds. These affinities "are sacred and they stretch back…to whole generations" (Marealle 1952: 60). Such totemic relations were not discernible during the fieldwork but there were many myths that involved specific animals and birds. For example birds are particularly revered for their supernatural communicative abilities. Owls for instance are noted for bringing messages – usually bad ones. Numerous informants told of premonitions instigated by repeated visits by owls (see MC/Ud6/July 2004; MR/Mb1/August 2004). Interestingly one interviewee, convinced that owls were communicating media for otherworldly beings, knew of some English folklore associated with wise owls. He believed that this reputation of wisdom was derived from their abilities to communicate (MC/fo32/July 2004). The crowned hornbill or *inguuma* is another bird associated with death. It is believed that if an *inguuma* (also sometimes referred to as *motutu*) cries whilst near a house the occupant will die soon after. So entrenched is this belief that most of those interviewed reported deaths after witnessing such events (e.g. MR/Ms14/August 2004). Others have noted that birds were a food adult men were forbidden to eat (Moore 1976: 360). This may relate to the special place birds have in the contemporary mythic world but participant observation suggests this does not apply to chickens and ducks.

In the Wachagga cosmology fortune and misfortune are brought on by im/proper ritualised conduct. *Ruwa* and remote spirits and ancestors control supernatural power. However, the superior spirits tended not to interfere with human affairs. The entities that exist in the spirit world, within the earth, need to be managed and appeased through ritualised human activity (Moore 1977: 46). Ritual action is known as *mrumo*. Activity considered *mrumo* consists of petitions, thanksgiving, reconciliation, libation, veneration and adulation. It has been commonly presumed that the Wachagga have a "dying culture and a dying language" (Marealle 1952: 63; see also Stahl 1965). Such postulation should be tempered with the fact that myth and ritual are still culturally important and seem to be offering resistance to the modification processes of urbanisation, capitalisation and centralisation.

> "[W]hile the superficial appearance of Chagga culture has all but disappeared, giving way to an image of accommodation and assimilation, the substratum of cultural values and beliefs continues, albeit in modified form" (Kerner 1995: 117)

The mountain still evokes some ritual action. In the 1950s it was commented that the elders continued to "look at Kibo with a contemplative silence" and think "we know your hidden powers will see us through this day" (Marealle 1952: 57). This relates to the older traditions whereby upon arising in the morning a Mchagga would face towards Kibo and commit prayer towards the mountain in praise (Marealle 1965: 57). Indeed such ritual behaviour has been proposed as contemporaneous (see Mosha 2000: 77) but it seems unlikely that even amongst the elders the spitting-rituals are practiced. Nevertheless the practice is still recognisable in living memory and as such coordinates experience and perception.

Much of the substance and inherent symbolism embedded in the precolonial ceremonies of the Wachagga (see Dundas 1968; Gutman 1909, 1926) survives in modified form in the mythic memoryscapes of the contemporary communities. Some examples can be posited concerning male initiation rites. Following circumcision the young Wachagga males were dispatched into the forest for a period of seclusion and education. The women and children were informed this was for the enactment of the anus stitching ritual. Moore (1976: 357) elaborates that this fictional operation was necessary because fertile men completely digested their food and so did not defecate. This closed anus fiction must, however, have been a open secret for male faeces was required in the female initiation rites (Moore 1976: 357, 361; Raum 1940: 350). Nonetheless to be closed was to be male and fertile. To be open was to be female and infertile. The male has a closed perineum and the female an open vagina. The term for anal plug is *ngoso* and it is

interesting that a pregnant woman is described as *ngoso* also (1976: 358). In short faeces are absorbed by virile men and menstrual blood is absorbed by pregnant women (1976: 366). Moreover the red soft components of the body are categorised as feminine and the white hard components masculine (Hasu 1999: 286). The closed/open duality dichotomised the cosmological world. Thus Moore (1976: 358) summarises:

(male-closed-virile) = (female-closed-pregnant)
(male-open-defecates) = (female-open-menstruates)

The male initiation ceremony also contained much other symbolism related to the anus and faeces. Essentially faecal symbolism, ingestion, anal stitching and seclusion were ritual manifestations of the rebirth into adulthood. The anal closure prevented another man becoming pregnant through homosexual liaisons because such activity becomes physically impossible – the initiate is figuratively closed to other men (1976: 359). The seclusion enclosure in the forest contained two pits – one for urination and one for defecation. The faecal pit was known as 'the bull' and was consecrated by rituals performed by the chief of the age-set and the *lodana* or supervisor of the ceremony. Perhaps it was called the bull because of the use of manure in increasing the fertility of the soil. The initiates were required to ritually consume animal meat and insects dipped in the faecal pit. This ritual doing of the forbidden marked the beginning of a lifelong prohibition that symbolically encompassed cannibalistic tendencies, homosexual contact, and consumption of taboo things (1976: 362). It is interesting to note and no doubt related that Wachagga men still treat locusts as forbidden food (1976: 362) and it is considered far ruder for a man rather than a woman or child to defecate in public (e.g. MC/Nr19/July 2004).

The Wachagga had ritualised forms of writing in the precolonial period. One example was a notched record that "represented figures and speech" (Marealle 1952: 59). The sticks were approximately 140-170cm in length and were divided into about twenty bark-ring sections. These bark-rings separated smaller carved-rings which exhibited notations in the form of vertical, inclined, horizontal and triangular scores (Kerner 1995: 118). There have recently been efforts by Wachagga intellectuals to investigate these *mregho* sticks (see Kerner 1995). The motivation behind such research agendas is to establish that a precolonial writing system, which articulated indigenous knowledge and culturally unified the tribal-ethnic group, existed (1995: 125). There has also been renewed interest amongst the young in the traditions of initiation which has been attributed to the changing political theatre and the AIDS pandemic (1995: 122-5). This interest has enhanced the remembrance of the past and highlighted its position within the memoryscape. The *mregho* are conceptualised as books with notches and inscriptions being chaptered lessons. These lessons are a corpus of ritual knowledge related to the initiation ceremonies. The *mregho* embodies memory – the relational memory of things and the memory of skills. These forms of memories are homogenised through the *mregho* stick and read as propositional knowledge i.e. memory about things (1995: 118). Propositional knowledge is temporary and detached from the object or person that possesses it. Kerner (1995: 126-7) provides an interesting remark made by a Mchagga elder about the *mregho* when considering their power which is "equal to that of the bible, both of which I keep under my bed to remind me of what I need to know".

It is still not known whether the notches are encoded in a standardised fashion or whether the notations were intelligible only to those involved in the related initiation ceremonies. It is likely that the *mregho* stick was a mnemonic of initiated instruction. *Mregho* instruction occurred after the male circumcision rites and the period of seclusion and recovery in the forest – with the age-set known as the *rika*. In contrast to the earlier stages of *rika*-initiation, which were conducted by the highest age-sets and under the purview of the chief, the *mregho* rites were performed by members of the next ascending age-set and by immediate relations (1995: 118). The *mregho* was performed.

> "The inscriptions on the stick were represented to initiates as a cosmology of the human body, and served as a schematic map that preserved the memory of how experience was to be interpreted and understood" (Kerner 1995: 118)

During the performance of the *mregho* rites *Ruwa* was evoked with the call "Stillness, oh Great Stillness" conveying respect whilst beckoning the silent listener (1995: 119). Lessons were given regarding the body, procreation, growth and morbidity. The rites were lifelong lessons and the individual would repeat them multiple times as required in different capacities i.e. initiate and initiator. The memories encoded in the *mregho* were historically modified through individual and communal interpretive, memorative and performative processes. The landscape is perhaps a macrocosm for the lessons and memories performed and embodied through the *mregho* stick rites. The landscape embodies cultural and religious knowledge that is accessed and performed during life.

PROTECTION, SECURITY AND SHELTER

The physicality of the mountain endows protection. *Uchagga* was the locale of a long tradition of political instability. In precolonial times warfare and raiding between rival patriclans and lineages was common. At the same time petty chiefdoms were campaigning with each other for local dominance (Dundas 1968: 285; Moore 1970: 325). This usually involved the warrior age-set performing raiding exercises for cattle, women for breeding, and men to be sold into slavery. The essence of this instability and fearful existence is captured with the following patronising remarks "these silly savages [could] think of nothing but mutual extermination" (Johnston 1886: 177). The negotiation of power relations through bloody force was habitually played out upon the *kihamba* land. It has been noted that the graphic traditions of battles, sieges and campaigns should not be taken too literally considering the small numbers of protagonists involved. The incidents are probably founded in actual events but the details have been embroidered (Fosbrooke 1954: 117). It has been estimated that petty chiefdoms in the precolonial period probably numbered between 500-2000 individuals (1954: 115). Nonetheless the exaggerating effects of memory and storytelling has brought about a memoryscape that stresses the scale of the bloodshed.

The chief's fortified their spheres of influence. Chiefly power enabled the requisition of labour for construction projects and the sequestering of materials for such projects (Moore 1970: 325). The most conspicuous of these defensive structures were/are the stone fortifications. There are numerous examples of such structures but the best preserved is that at Kibosho (Dundas 1968: 96; Fosbrooke 1954: 116-7; Wynn-Jones 1941: 11). Other forms of defensive construction were trenches and earthworks. Dundas (1968) mentions that the "country was secured by war trenches which were everywhere". Little evidence existed for these trenches in the 1950s (Fosbrooke 1954: 120) and none were identifiable during recent ground survey conducted around the three transects. The mountain was also protective in the provision of underground fastnesses. These were defensive shelters used to protect the local familial or patriclan group from other patriclans and other tribal-ethnic groups (Fosbrooke 1954: 115; Wynn Jones 1941: 11). These dug outs were located near to the dwellings usually within the 'ancestral bounds' (Wynn-Jones 1941: 11). During attack women and children would retreat to the refuge and remain there self-contained for prolonged periods of time. Some dug outs were comparatively small and would only be able to shelter a small number of people while others were more elaborate affairs which were designed to hold numerous people, heads of cattle, and incorporated underground water trenches and ventilation shafts (1941: 12). Some were so elaborate that they were an inter-generational inheritance and in fact some were never completed (Fosbrooke 1954: 126). They were constructed using simple hand-held tools resembling a hoe with blades smaller than 3cm in breadth (1954: 124). Indeed the hoe-marks can still be observed on the walls of some of the chambers in the Marangu bolt hole (1954: 124). This hoe-like instrument is also referred to elsewhere.

> "[T]his laborious engineering was being executed with the aid of a single tool in the form of a crow-bar…and with this ineffective implement they managed to bore through the

Figure 8. Photograph of Laban cave

volcanic rock immediately underlying the surface-soil" (Willoughby 1899: 214-5)

According to Wynn Jones (1941: 12) these hiding places were rarely discovered because the secret of their location was so vehemently guarded by the patriclan. It is clear that some of the bolt holes were utilised as some of the earliest oral traditions recorded incorporate these constructions in the unfolding narrative (Fosbrooke 1954: 126-7). Indeed in some chiefdoms it was an offence punishable by death to take refuge in the underground shelters rather than making efforts to repel the terrestrial attack (1954: 126). Moreover strategies to besiege and smoke-out those occupying underground fortifications were developed (1954: 127). There is even a Wachagga proverb that suggests the importance these structures were attributed '*manya ulamine upanga ulemuowa*', which roughly translates to 'do not neglect the cave that shelters you' (1954: 127).

Oral tradition and local knowledge recently permitted the location of two such bolt holes near the Machame transect. One such complex is locally referred to as the Laban caves (-3°11' 37°14'). Informants described how ancestral villagers would seek refuge in the caves during the wars between the rival patriclans of Kibosho and Machame. The caves are strategically well placed because they occupy a position below a ridge-line on a

riverine valley cliff and hence they are hidden from view from all directions. Visibility would have been further impeded because the whole area was undoubtedly heavily forested in the past (see Fernandes *et al* 1984; Fernandes and Nair 1986; Kivumbi and Newmark 1991). It should also be noted that the proximity of the river and waterfall to the caves would have been advantageous because noise would be disguised and any access/tracks hidden. Moreover spoil from the cave construction could have been deposited in the river and thus washed away or submerged. Another bolt hole was located based on information gained from interviews with elders at Foo village. The cave had not been seen by any of the informants but the living memory of their grandparents contained information about such things being in the locality (e.g. MC/Fo31/July 2004). The shelter is located in the Nkowoyo Kyalia area of Foo village (-3°12' 37°13'). The dug out is positioned on a steep rock-face and overgrown with thick vegetation. Various informants believed that the cave was one of a series of bolt-holds from the time of the patriclan wars.

> "Our ancestors would hide in that cave for safety while outsiders attacked. They would stay in the cave for many days. Animals and food had to be taken as well so that no one would have to risk leaving while the men would fight" (MC/Fo15/July 2004)

The informant's remarks perhaps highlight how memories become exaggerated and confused. The Laban caves (**figure 8**) would certainly have been able to provide refuge for many people and the considerable supplies necessary to withstand a prolonged besiegement but even allowing for the accumulation of detritus, subsidence and/or partial collapse the bolt hole at Nkowoyo Kyalia (**figure 9**) would probably have been too small for multiple people and goods. Preliminary inspection suggests that rather than being a natural feature the bolt hole was bored or at least partly integrated through modification of the volcanic rock. There is evidence of instrument-made markings within the cave similar to those identified at another Wachagga underground complex (see Fosbrooke 1954: 124) and in an example from Northern Pare (Fosbrooke 1935: 5). Interestingly there is some evidence that the Wachagga may have been contracted to construct the Pare bolt holes and tunnels (Fosbrooke 1935: 5). These caves feature in the mythic environment of the local populations with the sense that protection, refuge and safety is integral to all perceptual knowledge of the mountain.

Figure 9. Photograph of Nkowoyo Kyalia bolt hole

CHAPTER SIX

MOUNTAIN BOUNTY: THE LANDSCAPE AS PROVIDER

PROVISIONS

The landscape of *Uchagga* is culturally understood as fertile and (re)productive with many facets to such provision. Wachagga land tenure notes the existence of *shamba* and *kihamba*. *Kihamba* land is the home or dwelling of the Wachagga. It is where the house is constructed, where the bananas are grown, where the ancestral spirits reside, and where mothering, education and growing takes place. It can be expressed that "the heart of the Chagga is in the kihamba" (Johnston 1946: 12). The preferred style of house underwent a huge modification over the last century. The traditional *msonge* has been replaced with the new bungalow-style house constructed of concrete blocks, sheet metal and other locally produced materials (Bailey 1968: 163). These obviously range considerably in value, quality, sophistication and standards. A few *msonge* are still erected to satisfy tourist interest and to accommodate heads of cattle. Examples of these dwellings are shown in **figures 10.1, 10.2 and 11**. This modernisation of the architectural landscape accompanied the indigenisation of Christianity (cf Comaroff and Comaroff 1997: 277-322).

The *kihamba* could be considered permanent freehold property (Johnston 1946: 1). The relationship between a patriclan and particular places or plots of land has been documented. Johnston (1946: 5) notes the rare circumstances in which a *kihamba*-owner could be removed of his right to the land historically and toward the end of the British administration which included banishment, prolonged absence or non-occupation, or the allocation of *kihamba* land in a neighbouring chiefdom. Yet even in these circumstances the dispossessed would usually be able to claim back the land through the native courts (1946: 5). This highlights the close relationship individual and patriclan identities have with particular places and how rights of tenure and habituation correlate with past agency, histories and memories.

"Kihamba land is very important. It is very fertile land and such areas give good harvest. The land has and will always be good to my family. My sons will eat the crops from here and then in turn so will their sons. We have always fought to keep hold of this land" (MH/Mk35/September 2004).

In contrast *shamba* land is found on the lower slopes and is usually used to farm maize. It used to be annually allocated by the area chief on the basis of temporary purposes and needs (Johnston 1946: 1). However, most of the former *shamba* land has been turned into *kihamba* land due to the acute shortages and inheritance rights. The inheritance of land has traditionally been a complex affair (see Johnston 1946). Upon the death of the father the eldest son becomes head of the family. The youngest son is the privileged heir in that he will usually inherit most land (1946: 7). The middle sons or *mrina* have no cultural inheritance rights and will usually seek to establish new holdings or find paid employment (Bailey 1968: 165). It is also the case that bush or *pori* and the forest belt are still sometimes referred to as *shamba*. The soil on Kilimanjaro varies from the humic brown soils of the forest and banana-belt to the reddish loams found at the lower altitude (see Holand 1996: 36). The soil is of volcanic origin and nutritiously rich (Hermansen *et al* 1985; see also Mnkeni 1992). There is some evidence that the irrigated soils are beginning to suffer from processes of leaching, sodicity and salination (Tagseth 2003: 21). Various agro-ecological zones have been proposed on the mountain but the subtle differences are largely irrelevant here (e.g. see Conyers *et al* 1970; Hermansen *et al* 1985; O'Ktingati and Kessy 1991; Tagseth 2003). The cultural concepts of *shamba* and *kihamba* land are still apparent although the system of land tenure is perhaps beginning to be somewhat revolutionised with the origins of larger-scale farmsteads and agro-industry especially in the lower-belt land of West Kilimanjaro. The fertility of the soil in the *kihamba* is declining as well. One informant in Tagseth's (2003: 62) interviews remarked "the soil got tired" after lamenting that in the past small plots of land were highly productive. The provision of the mountain is failing to sustain the growing population and poverty and malnutrition are commonplace (see Howard and Millard 1997). Nonetheless the mountain is still perceived as the great provider.

Crops are of course not the only provisions of the mountainous terrain. As one respondent noted "water starts at Mawenzi peak and on its way down it brings life; it also gives good air, fire wood, good climate and rain for life" (MR/Ms8/August 2004). The mountain also provides protection from enemies and wild animals and therefore brings "peace and stability" (MR/Ms21/August 2004)). Some interviewees even commented that it provides babies (e.g. MR/Ms8/August 2004; MC/Fo11/July 2004; MH/Mh2/August 2004). The mountain is also believed to have created the waterfalls, lakes and forests. Traditionally cultivation of the staples *mbege* (eleusine/millet) and *mrarao* or *mkonasi* bananas takes place (Bailey 1968; Curry 1938; Johnston 1946; Swynnerton 1949). Various cash crops have been introduced and intensively farmed including coffee and sugar cane (Bailey 1968; O'kting'ati and Kessy 1991). The mountain also provides moral strength and watches over the Wachagga. Thus the mountain is the greatest provider. During the fieldwork whenever questions were asked about why the Wachagga came to live on the mountain the answers always made reference to the provision and exploitability of the landscape in terms of climate, air, rain, water, soil, game, safety and so forth. Generally those that believed the Wachagga had been put on the slopes of Kilimanjaro by *Ruwa* thought that the provisions were other-worldly reward and those that believed that the Wachagga were of mixed tribal descent from all over East Africa thought that this place was ideal for settlement owing to the ample resource provision.

The multi-origin model is the popular academic interpretation (see Stahl 1965; Odner 197) and it is also the tribal origin-narrative taught at all the local secondary schools in the Marangu district. It is very important to note that virtually all interviewees reported that the mountain was important to them and their patriclan because it gave them rain which seems logical considering cultivation is the primary mode of subsistence. There is an important distinction to make regarding these comments. This instigative relation between the mountain and rainfall did not necessarily relate to the 'water cycle' a familiar geo-climatological abstraction of the modern West, but to a more fundamental causal link. The mountain, as a personified entity, provisions rainfall for the Wachagga – for the cultivation and nourishment requirements of their daily existence.

Figure 10.1. Photograph of a traditional msonge constructed with banana leaves

Figure 10.2. Photograph of a traditional msonge constructed with grasses

Figure 11. Photograph of a modern dwelling sited in the kihamba

This sense of ancestral or supernaturally-ordained provision caused considerable friction between the locals and the administrators of the Kilimanjaro Forest Reserve and later the Kilimanjaro National Park during the colonial and postcolonial administrations. This can be highlighted by a letter to the Provincial Forest Officer from the Conservator of Forests, Northern Circle in 1956 when concerns regarding the 'half-mile strip' were rife.

> "[D]emand by the local people for building materials and fuel has led to unchecked over-exploitation of the lower half-mile or so of the forest to such an extent that a serious shortage of forest produce had come about" (TNA/KR/10/B/7/GEN/G)

Essentially a poor fertility-zone had been created. Recently this area has in effect been returned to the locals in the form of the village and local government fuel belt. Within this belt no farming is permitted as it may give rise to the claim of title to the land by the farmer. In areas where the village and government fuel belt does not exist some farming practice is exhibited. There is a fundamental – traditional and contemporary – discord between centralised government and the Wachagga. For

instance it has, since the colonial era, been proposed that "all forest crops belong to the state" (TNA/KR/15). It was recognised though that the role of the mountain forestry organisations should be protective and not productive. Nevertheless attempts have been made to cultivate wattle, pyrethrum trees and softwoods in the Marangu half-mile strip. The Wachagga believe that the forest produce should belong to them or certainly that the fertile zones should be theirs to exploit. This dissatisfaction is most noticeable in the elders but recent concerns about water provision and access rights are similarly contested by all sections of village communities (see Tagseth 2003 for more detailed discussion).

Despite the historical provision of the mountain environment one is still expected to pray for its continuation and amplification as the head of one household commented "if I pray through the mountain I get what I need" (MR/Ms37/August 2004). It is also interesting to note that the notion of mountain as provider and nourisher is also related to Christianity. One woman noted

> "The mountain is good to me because it brought missionaries. Thus the mountain gave

43

me education, hospitals and their [the missionaries] culture. It helped me see God in the new ways" (MR/As14/ July 2004)

Thus the mountain is conceptualised as the initiating force behind the arrival and continued presence of the missionaries and the church. This is another example of the missionary teachings being made intelligible through cultural perception and coherence. In a sense the high concentration of inhabitants on the mountain did precipitate the arrival of the missionaries but this is clearly not the meaning of the interviewees.

WATER AND IRRIGATION FURROWS

Kibo's glacial peak provides a seemingly inexhaustible supply of water that through streams has cut deep gorges into the slopes. These gorges have developed into natural boundaries on the landscape that related to the borders of the historical chiefdoms. In fact many of the contemporary ward and divisional boundaries relate to these also (Moore 1970: 325). Water and its vitality hold other cultural importance. Water is associated with life-bring and death-bringing and as such it possesses supernatural potency. A river near Marangu is actually called *Ruwa*. Water can be angry and dangerous but it is also conceptualised as inherently positive.

> "It is water that can put out the fire that kills, water that can quench death-bringing thirst, water that is benign and cooling, water that …can be used to bless and purify and make all dangerous things peaceful" (Moore 1976: 365).

In its positive thirst-quenching and nourishing role water is symbolically associated with both milk and blood. Female milk feeds children, cow milk feeds all, and male milk feeds the vaginal mouth (Moore 1976: 365). Milk like water is thus life-bringing. The Wachagga also conceptualise blood as life-giving. The blood of the slaughter is linked with sustenance and nourishment and the female blood, as the recipient of male milk, makes children (1976: 365). In the Wachagga cosmological world blood can also be death-bringing. For example the male blood of warfare and the menstrual blood of a non-pregnant woman (Moore 1976: 365). Blood is also one of the products and prerequisites of growth. The preparatory blood of male and female circumcision is a necessary shedding involved in processes of maturation. Water is also conceptualised in such bloody terms.

> "The Chagga stay on the mountain because it gives us water. The river and *mifongo* systems sustain us, our cattle and our crops. Without the flowing water we would all die. This is why we love the mountain – it provides us with what we need and keeps us alive" (MR/Ly4/August 2004)

Uchagga is home to one of the most advanced and complex system of furrow irrigation in the world (Gillingham 1997, 1997; Grove 1993; Pike 1965). The furrows have had a long history of indigenous regulation. To this day there is a complex administrative system involving local furrow boards and furrow elders that have recently seen much tension and dispute with elected water user groups and more centralised water regulators (Tagseth 2003). The furrow method of abstraction has not been extensively altered since precolonial times (Masao 1974: 1; Pike 1965: 96). A former water warden on Kilimanjaro noted that the furrow system reflected favourably on the ingenuity and industry of the indigenous population in maximising the agricultural potential of the area (Pike 1965: 96). Furrow-cutting is a highly specialised art that has virtually died out. The existing furrow schemes are well maintained and utilised (Tagseth 2003: 92). In the past furrow-cutting was a male enterprise and the oral tradition reported in Masao (1974: 2) contends that only certain patriclans possessed the specialised knowledge to perform such cutting - such as the Wakomfongo. All work associated with the furrows is still within the male labour sphere (Tagseth 2003: 92).

It is reported that the cutting of new furrow was a highly ritualised affair. The plans of such enterprises were either motivated through dreams or necessity (Masao 1974: 2). The project was ordained by the local chief and a period of prayer undertook. The petitions were directed to a recently deceased elder who it was assumed would entreat with the more remote ancestors who were hierarchically privileged in the scheme of the supernatural. There is a myth that termites or 'red ants' were the chosen physical means whereby the ancestors colluded in the furrow-cutting endeavour. The termites would march in single file through the area to establish the necessary route of the furrow (Masao 1974: 3; Stahl 1964: 212). Masao (1974: 3) notes that termites are particularly adept at avoiding obstacles and thus there may be some exaggerated truth to this legend. Repairs and modifications to the furrow are also observed to require ritual action. Traditionally child sacrifices were drowned at the intake (Dundas 1968: 263), similarly the drowning of heads of cattle has also been deemed appropriate (Masao 1974: 5). If the furrow dried up or diminished in productivity appeasement of the spirit world was required (1974: 5).

The irrigation furrows embody a religious dimension. The Wachagga developed intensive agricultural practices about four hundred years ago, which was precipitated by the narrowness of their preferred ecological niche (Tagseth 2003: 14). The early colonialists noted elements of this intensive agricultural practice including zero grazing (Johnston 1886), water canal construction (Rebmann 1860), furrow surveying and appreciation of grade (Raum 1940), and high-density irrigative furrow systems (New 1873). Furrows are of great importance to the indigenous populations they supply and this is not simply for agricultural and productive reasons. One administrator of Moshi during the British Colonial Administration noted that "furrow rights are sacred" (Dundas 1924: 264). Especially important are the furrow intakes. Interference with the intakes, even unauthorised approach, can be interpreted as sacrilege. Indeed in some areas purification ceremonies are required on such occasions (Tagseth 2003: 81). There has been a long history of water-access disputes involving the Wachagga (see Mung'ong'o 1997). The Wachagga perceive the water as a local resource to which they have prescriptive access. These rights have been ordained by history and tradition.

> "The water does not belong to the water office. It is given to us by God. Our ancestors took water from this river. To pay for it would be a sin" (Tagseth 2003: 80)

With higher water demand from downstream communities, higher population densities upstream, and growing requirements for electricity in the urban centres these access rights are bound to cause friction in the future. The nourishing arterial waterways may soon be clamped with limitations on water usage and access restrictions enforced by centralised authorities.

Irrigation furrows possess a sacred quality and this trait was highly visible on all three transects of the study. Indeed not only is the furrow itself described in sacred terms but also the person who historically initiated its cutting. These sacred ancestors inhabit the memory of the communities that reap the benefits of the water provision.

> "The headman Mkawe who went up the mountain and dug the furrows to Maharo was a great man. He blessed us with the water three generations ago – without him the land would be too infertile and dry to support us" (MH/Mh61/ September 2004)

The primary furrows that feed Maharo and Makiidi villages derives from a body of water higher up the mountain known as *Nduwa*. This entity frequently undergoes processes of personification. *Nduwa* must be respected, treated with deference and paid homage. The body of water can become angry and on such occasions the water will cease to flow. During such times animal slaughters must be enacted in order to placate such antagonistic tendencies (MH/Mh3/August 2004; MH/Mk40/September 2004). In order to abstract water from *Nduwa* you must, after enlisting the support of other furrow committees and elders, perform the ritual sacrifice of two black sheep or goats. One should be drowned in the body of water close to the abstraction point or intake and the corpse sunk and the other should have its neck cut. The second animal can be consumed but this has to happen in close proximity to the proposed intake. The remains should also be discarded into the water. This double slaughter has to be repeated

every five years to keep the water flowing (MH/Mh34/August 2004). It is commonly thought that the current water shortages in the furrows have been caused by the failure of the furrow-diggers to uphold this practice in the presence and influence of the world religions.

> "How can they [other furrow users] expect the water to flow when we do nothing to keep it coming? Our ancestors knew how to live here and had plenty of water because they respected the traditions and customs of the land" (MH/Mk27/August 2004)

Currently the practice of secondary furrow digging is a controversial one in the Maharo area. The local water regulatory bodies and government agencies want the practice to end but there is also contention amongst the locals who recognise that the water is in finite supply (MH/Mh3/August 2004; MH/Mh17/August 2004).

It has recently been pointed out that the intensive furrow-irrigation system of the Pare is very similar to that of the Wachagga (Sheridan 2002: 82). Therefore some of the symbolic relations may well relate to the Kilimanjaro region also. The Pare dwell on the Northern Pare mountains and underwent agricultural intensification in similar low population conditions to those witnessed in *Uchagga* (Håkansson 1995, 1998). The Pare case study is a relational analogue because similar ideological forces and environmental pressures are discernible. The Pare and the Wachagga irrigation systems both predate the colonial administrations and most furrows are still operational. The Pare recognised two irrigation intakes. The *irombo* are river or spring intakes consisting of organic matter, clay and stones that divert part of the riverine flow into the furrow (Sheridan 2002: 86). The *ndiva* are reservoir intakes constructed in a semi-circular arrangement below a spring these intakes structures are covered with wood and an outlet at the base or *lukotho* is regulated by a *ngoso* plug (2002: 86). Establishing new intakes required sacrifices to ensure ancestral blessings. During the field study in *Uchagga* the author did not encounter any irrigation constructions similar to the *ndiva* but the *irombo*-type intakes appear cross-cultural. It is proposed that the symbolic dimensions may in some fashion translate also.

Irrigations systems are forms of symbolic and material capital that can only be understood through analysis of both ideology and technology (Sheridan 2002: 79). The shared landscapes of the Pare and the Wachagga were inscribed with dominant ideologies of gender inequality and peripheralisation. This is exhibited in the preventative taboo that regulated female access to irrigation intakes despite the fact they are historically and contemporaneously responsible for the majority of agricultural labour (2002: 82). It has been reasoned that this taboo relates to fears about menstruating women upsetting the life-giving flow (2002: 88). Menstruation is linked with sterility and death in the Wachagga cosmological world (Moore 1976: 358).

There was much other commonality in ritual practice in precolonial communities neighbouring *Uchagga*. Pare initiation also somewhat mirrors that of the Wacchagga. The initiation rituals and institutionalisation were conducted in seclusion in sacred forests (Sheridan 2002: 82). The Pare had similarly engineered artificial bolt holes excavated into the mountain for protection (Fosbrooke 1935). It has been proposed that the techniques used in constructing the Pare dug outs originated from *Uchagga* (Fosbrooke 1954: 115). Moreover the patriclans of the Pare are known to keep the skulls of their ancestors in sacred forests (Sheridan 2002: 89). The *Dracaena usambarensis* plant was utilised in the demarcation of land boundaries. The Pare/Chasu term for this plant is *isae* (2002: 85). The furrows and irrigated landscape are charged with sexual symbolism. It has been noted that *lukotho* literally translates as 'that which pours forth' but it is also a slang term for the phallus (2002: 88). Moreover the closed/open dualism is readily discernible in the Pare cosmological world – in fact the Wachagga instrument of anal closure the *ngoso* is also the instrument of closure that keeps a *ndiva* impregnated with water. The replication of the *ngoso* term may undoubtedly relate to a common philological origin. Nevertheless the high rates of migration and cultural exchange between the Pare and the Wachagga favour the conclusion that the symbolic dimensions were intra-cultural. Clearly the life-sustaining water and its inscription upon the landscape in the form of furrows are extremely important features in the Wachagga symbolic and cosmological order.

GEOPHAGIC CONSUMPTION OF SOIL

Geophagy is the consumption of earth/soil and is apparent in many cultures (see Abrahams and Parsons 1996; Andrea and Katz 1998; Poda *et al* 1999). The practice of geophagy has been documented in the Kilimanjaro region in the past (Knudsen 2002). It has been noted that geophagy is a form of indigenous medical knowledge that is conducted covertly due to the normativity established by Western white/missionary medical practice (2002: 24). Indigenous medicine encompasses methods of healing, prevention and prophylaxis, and the maintenance of well-being for both individuals and communities (2002: 22). It has been noted that while one school of thought regards geophagy as pathological behaviour (e.g. Horner *et al* 1991) another notes some intrinsic medical value to it that is based on normative Western notions (e.g. anti-diarrhoeal effect Vermeer and Ferrell 1985; and detoxification effect Johns and Duquette 1991), whilst a third notes the value of the practice in terms of cultural significance (e.g. Knudsen 2002). The cultural significance of geophagic practice is of considerable

interest to the investigation of spiritual dimensions to the landscape. Indeed Knudsen notes that *akula udongo* or geophagy is "embedded in ways of being woman and in cultural symbolism" (Knudsen 2002: 20). The practice is dependent on the psychological and bodily needs, tastes, and social responsiveness of individual women and their related communities. The appetite for, and satisfaction from, soil-eating is conceptualised and articulated as comparable to sexual intercourse (2002: 22).

> "The women revealed they had experienced 'Tamaa' [desire] and consumption of soil in some pregnancies and not in others. The women made a distinction between hunger for food and hunger for soil…The women expressed specific preferences for different forms of soil. The soil also had to be obtained from *specific places*, and was either preferred dry or wet." (Knudsen 2002: 20, my emphasis)

Wachagga geophagic practice is embedded in cultural symbolism that further relates one's identity to the landscape. The woman's identity is closely intertwined with the productivity of her husband's *kihamba* land. The productivity of the land is linked to human reproduction. This geophagic landscape symbolism has a connection with termites. The household is tied to the woman's identity. '*Nyumba ni mwanamke*' or 'the house is the woman' is a well-known proverb throughout *Uchagga*. The termite queen is conceptualised as an idealised mother and model of femininity. Whilst pregnant a woman should reinforce her association with the termite queen (Knudsen 2002: 22). Thus she should be fertile and life-giving but also remain responsible for production in the homestead. One means to reinforce the symbolic feminine-bond with the termite queen is to consume soil from the *kichuguu* or termite hill. The termite hill is believed to provide emotional and bodily well-being (2002: 20). In Knudsen's (2002: 23) study over 90% of the sample consumed soil from termite hills.

It has been noted that Wachagga artefacts that have a symbolic dimension related to magic, life-enhancement and spirituality are often decorated with depictions of sexual organs – the agencies of life and death (Knudsen 2002: 22). The soil generally and particular places specifically are appreciated and bodily understood as sexually desirable and satiating. Indeed the earth itself through its geophagic and agricultural productivity and life-sustaining properties is spiritual. Moreover the earth is a macrocosm for the smaller pieces of Wachagga material culture, which are alluded to by Knudsen, that exhibit spiritual and magical functionality. Some of the symbolic associations of geophagy and reproduction are depicted in **figure 12**. It should be noted that the associations are depicted vertically as well as horizontally.

Figure 12. Symbolic dimensions and associations of geophagic and reproductive processes

Soil	↔	Seed	↔	Nourishment	↔	Crops
Termite Hill	↔	Queen Termite	↔	Mothering	↔	Reproduction
Vaginal Seeding	↔	Pregnancy	↔	Geophagy	↔	Birth
Birth	↔	Geophagy	↔	Education	↔	Adulthood

The soil is a medium facilitating the transference of the spiritual properties of the landscape to the individual. Moreover the geophagy unites and bonds the individual, the patriclan, and the tribal-ethnic group with the landscape.

> "We eat the red soil when we need to because it makes our babies belong to the mountain, it makes them strong like the land, and it means they will love the land. This is important to the women in our clan" (MR/As28/July 2004)

Primarily the practice concerns pregnant women but children are also known to partake in geophagic consumption and, of course, unborn babies share in the life-giving character of the ritualistic consumption (Knudsen 2002: 19). During the fieldwork venture all three transects identified individuals who partook in geophagy with the practice being most widespread amongst the Marangu transect respondents. Nonetheless it must be acknowledged that this discerned universality of soil consumption in Marangu may relate to the practice being less covertly practiced there rather than it necessarily occurring with greater frequency.

TOURIST PILGRIMAGES

The mountain was frequently referred to as the 'Old God' during the interviews along with other terms that demonstrate an acknowledgement of potency and sacredness such as 'Sleeping Giant (e.g. MR/Ms22/August 2004)', 'Old Man' (e.g. MH/Mh57/September 2004), and the 'Highest Land' (e.g. MC/Nr8/July 2004). It was also commonly believed that the constant flow of inbound tourists were being attracted to engage the mountain's potency rather than merely arriving to view or climb their mountain devoid of supernatural significance. In this sense the tourist influx is conceptualised as a form of mass pilgrimage. The elders believed the National Park boundaries were restrictive and prohibitive in that they stopped people getting "close to the wonders" (MH/Mh12/August 2004). The National Park is recognised as a form of administrative control involved in the regulation of power. Boundaries regulating access to the mountain have traditionally been contested. Indeed during the colonial period the local administration had a difficult time enacting policy. For example, the reaction to the establishment of Kilimanjaro Forest Reserve was violent in places – especially Marangu and Mwika. Indeed the boundary pillars were so frequently vandalised that extreme penalties had to be sanctioned. Punishments included a 3000 shilling fine, extreme when one considers a chief's wage was 500 shillings per annum, and/or six months imprisonment. A letter from the Divisional Forest Officer, Northern Province to the District Commissioner in Moshi noted the harsh punishments awarded for "defacing, injuring or removing any forest boundary marker should be impressed upon the people of Kilimanjaro" (TNA/KR/22/8/1). A further problem was illicit grazing within the reserve by, amongst others, Wachagga squatters that catalysed the implementation of forest guards and patrolmen (see TNA/KR/1953/8).

The mountain is seen as attractive in that it proactively pulls things and people towards itself. This supernatural force coordinates the environment around the mountain. The term 'magnetism' was frequently used to articulate this attraction (e.g. MH/Mh4/July 2004). This force of attraction is how the indigenes explain their presence, in-migration, tourists, and other phenomenon. This power has also been deemed responsible for the numerous plane crashes on the slopes. Indeed numerous tales of plane crashes were reported during the fieldwork and explained as being instigated by *Ruwa*. Planes that try to cross the mountain are trying to look at and compete with the divine and therefore are destroyed. There are obvious parallels between this myth and the biblical tale of the Tower of Babel.

> "The planes and those people they try and look at the mountain and see what is up there. God does not like this prying into His business and they are made to fall from the sky. Most of the planes you can see in the forests where they hit the ground" (MH/Mk18/August 2004)

This force of attraction is a generic quality of the mountain but it is also specific to certain places. For example stones that one might recognise as 'slippery' are thought to possess *mbarimu* that will pull you over and cause you to lose balance (MH/Mh19/August 2004; MH/Mh29/August 2004).

LOCAL ENVIRONMENTALISM

Due to increasing population density competition for land and resources has precipitated government initiatives designed to impede destructive farming and subsistence practices. The case study material provided below relates to the Machame transect but similar initiatives and schemes are also found at the Marangu and the Maharo areas. In Machame such programmes have been successfully taken-up despite the inconvenience and burdens such action entails. This environmentalism has 'rung true' in the community and there is a sense of a religious dimension to such considerations. Moreover environmentalist discourse regarding pollution, contamination and other destructive processes have been culturally augmented. For instance, the lack of rainfall is explained by noting "up at the mountain there is plenty of dirtiness that makes the rain not fall and the streams run dry" (MC/Fo15/July 2004). This probably relates to confused notions concerning global warming and water aquifer pollution. These notions have undergone processes of 'cultural logic' and thus somebody's abuse of the mountain results in deprivation of nourishment. Such reasoning is probably enhanced by religious concepts like desecration and sacrilege that permeate through the communities through their dissemination from the missions, schools and churches.

A good case study concerning recent environmental work in the area is *Foo Development Association* (FODA). FODA is a Non-Governmental Organisation engaged in various activities concerning conservation and environmental sustainability. In this capacity FODA is strongly linked to the *Tanzania Traditional Energy Development and Environment Organisation* (TaTEDO) who source local start-up and establishment costs. FODA's mission statement reads:

> "To empower the people in order to revive, sustain, propose, and improve economy, education, health, socio-cultural environment and infrastructure of villagers by involving intellectual, professional and entrepreneurial contributions aimed at solving the challenges of targeted priorities" (FODA promotional material)

One of these targeted priorities is environmental security. Newly developed technologies based on traditional techniques are being introduced including rainwater harvesting, sand filter purification of water, drip irrigation, ventilated pit latrines and the production of biogas. Furthermore methods of sun and solar drying are being actively encouraged. Solar power systems are being promoted throughout the area and low-impact fuel consumption (see **figure 13**) economically and educationally supported. It is also noticeable that efforts are being made to ensure the communication of appropriate energy technologies and best practice demonstration based on indigenous technical knowledge.

It is interesting that colonial discourse characterised traditional production as backward and infantile (Yngstrom 2003: 177). Moreover it has been noted that colonial and postcolonial administrations related environmental improvement with modernity, and modernity was posited as a masculine process (2003: 175). In the Kilimanjaro region there are some appealing developments and projects that seem counterintuitive to such established doctrine. For example, the fuel-efficient technology cookers sold through FODA are advertised and marketed to women. Being modern is the enterprise of women and men alike. Local posters read '*Kuwa wa Kisasa*' [be modern] and '*tumia jiko sanifu la kuni*' [by using modern cookers] whilst accompanying photographs depict women utilising this state-of-the-art apparatus. According to such promotional material women are the prime movers in the modernisation process.

It is clear that the Wachagga have a complex relationship with the mountain environment. Traditionally and contemporaneously it is seen as protective and nourishing – concepts not dissimilar to the gods of Judeo-Christianity and Islam. The landscape is protective in that *pori* animals are infrequently encountered. It also historically offered protection for the Wachagga. During the tribal and patriclan wars (see Dundas 1968; Fosbrooke 1954) the mountain offered refuge and hiding places. There is also the notion that the mountain protects spiritually as well through ancestral forces and other supernatural agency.

> "Kibo and Mawenzi protect our clan from misfortune. They remind us how to behave and to be good and true at all times. We know that they can punish us for doing bad things like hurting our neighbours or polluting the water we drink" (MH/Mk37/September 2004)

The mountain nourishes through its climate and fertility. It provides plentiful amounts of rainfall, water channels (both natural and human-made), light, warmth, fertile soils, grazing for livestock, forests for firewood, bee-keeping, game-hunting and medicines. Tourism is also seen as a further form of economic nourishment provided by the mountain.

Figure 13. Photograph of environmental cooking technologies from the Machame area

CHAPTER SEVEN

RESPECTING THE PAST: ELDERS AND ANCESTORS

ORIGINS OF PEOPLE AND THE WORLD

The Wachagga have complex relations with the landscape. What of their traditions of origin? As Vansina (1985: 21) notes all communities have a representation for the beginning of the temporal order, humankind or the world that references their particular sociality. Wachagga mythology notes that the world has always existed (Dundas 1968: 107; Tagseth 2003: 14). Origin myths thus relate specifically to human communities and the mountain. Mount Kilimanjaro or 'Kibo' is the finite focus of the Wachagga identity (Stahl 1964: 20). Kibo is the Wachagga name for the mountain meaning 'speckled'. There are two dominant myths that account for the formative occurrence of Kibo – the stories of Tone and the separation of the Kibo and Mawenzi peaks. Both of these myths involve ancestral time beings that dwelled on the landscape. The legend of Tone concerns an ancestral individual 'Tone' who was banished from the community of humankind and consequently lived in the forest amongst spiritual entities. The supernatural cattle under his guardianship were responsible for regurgitating the mountain and related features (Dundas 1968: 33-5). The separation myth concerns the sisters Kibo and Mawenzi who fought over food stocks. Kibo was alleged to have taken "a big ladle, and aimed a damaging blow on Mawenzi's back, which gave her the rugged appearance that she still bears" (Dundas 1968: 36; Marealle 1952: 58). The myths that explain the origins of human communities also relate to the mountain. Recorded oral traditions vary but a common theme amongst a large proportion of Wachagga was the belief that they came from the mountain or were "dropped there" (see Stahl 1965: 37; Odner 1971: 132). Furthermore according to some religious traditions of the Wachagga, humankind were created and descended to earth from the sky (Lema 2002: 44). Dundas (1968: 43) proposes that 256 of the 732 clans on Kilimanjaro share this origin belief. This made this oral tradition by far the most prevalent. The Wachagga origin stories describe a process of migrating and settling in a general downward direction. Furthermore oral histories provide the information that the Wachagga have lived on the mountain for approximately 450 years. It is worthy of note in such discussion that the Wachagga history described in the oral patriclan traditions that can be corroborated with European documentation are meticulously complete and accurate (Stahl 1965: 37). Moreover the "continuous living past" of the Wachagga pervades all aspects of their existence. It would seem fair to presume high degrees of precision in the oral traditions that stretch further back.

Recent archaeological evaluation of the Kilimanjaro area is notably absent. Nevertheless some work has been done in the locale (most notably Fosbrooke and Sassoon 1965; Odner 1971; and Mturi 1986). Some of these works suggest a different origin for the Wachagga. Odner (1971: 145-8) concludes that the peoples of Kilimanjaro originated from early first millennium AD "Bantu-speaking" agriculturalists which were also responsible for the colonisation of other areas of northern Tanzania. Prior to this migration there may have been other inhabitants of the area. It has been proposed that these were most likely Wakonyingo or Wateremba pygmies (Moore 1977: 5; Stahl 1964: 37). These were probably assimilated as the Wachagga have been documented to be an absorbing people (Stahl 1965: 36). Oral traditions highlight the presence of ancestral paths and temporary stopping places over and to the mountain from many directions which strengthens the migratory origin models (see Wimmelbücker 2003). Wimmelbücker (2003: 44) also proposes that the bulk of immigrants came from the east and south hence the lower population densities seen on the western slopes. There is also some evidence of 'pastoral Neolithic' communities dating 1500-5000 BP. It has been suggested that some kind of agricultural practice may have occurred alongside pastoralism in these pre-Iron Age populations (Mturi 1986: 63). The earlier colonisation date of 5000 BP has been corroborated with C14 dating of anthropic material (Pomel 1999). The archaeological evidence suggests the possibility that there has been a continuous occupation of Kilimanjaro since these pastoral Neolithic populations – although admittedly the picture is far from complete. The fragmentary record is even more inopportune for the eastern and southern slopes.

So where did the Wachagga originate from? As noted some literature highlights the patriclan traditions that the Wachagga came downwards 'from' the mountain. Indeed there used to be the belief that the Wachagga came directly from the mountain. One famous myth based on this understanding notes, as described by Dundas (1968: 108), that *Ruwa* liberated mankind by smashing a vessel in which the first humans were imprisoned scattering them over the mountainside. This sits in contradiction to recent historical accounts that propose the settlement pattern was in an upward migratory direction due to the unfavourable agricultural conditions found at altitude (see Wimmelbücker 2003: 46-7). Nevertheless preliminary appraisal of past work suggests this may be the case. Odner (1971: 134) notes in his survey that there were more sites in the upper and middle zones than the lower zone. If the Wachagga did come 'from' the mountain then this finding provides insights into their precolonial religious beliefs and their persisting origin myths. It is certainly the case that the lower areas were seen as undesirable in earlier times (Odner 1971: 134). This may have been due to hostile contact with the Masai, water shortage, and the high incidence of tsetse fly and malaria (Brewin 1965: 115). The historical accounts give some credence to the claims of the indigenous educated elite during the British colonial administration who claim the tribe-ethnic group originated from the Akamba, Teita, Masai, Pare, Sambaa, Kahe, Kwavi and Dorobo (Marealle 1952: 58). Since the European colonial period the lower land has been utilised but the favoured agricultural ground has historically been a highland belt lying within the altitude 2000-3500 metres above sea level. The oral narratives perhaps allude to how the mountain landscape has become so inextricably linked to the Wachagga identity that it subconsciously features in their conception of origin. Clearly landscape memory adopts fictional elements and narratives within its character.

ELDERS, PATRICLAN RELATIONS AND LEARNING

A Mchagga elder or *msongoru* is an individual of age and experience. One can also be an elder through the occupation of a position of leadership – the chief is thus the most revered human elder (Mosha 2000: 42). Patriclan elders are the most respected members of society. It has been claimed that patriclan membership is "a sacred obligation" that "remains perhaps the only thing that can control the younger generation" (Marealle 1952: 59). There is a sense that an individual cannot exist devoid of relations to other patriclan members. This is highlighted in the proverb '*molaa tengonyi nyi mooru o saro una kigoro kii wanda, mbororo tsiknenda kwi*' or 'he who sleeps in a small hut behind the main building is just like a beehive with unknown contents' (Marealle 1965: 57). A more metaphoric and less literal translation would propose that 'one should not marry an individual without knowing their relations – they may have problems that may contaminate your clan'. The communality of the patriclan takes primacy over individual agency. A Mchagga is bonded to the history, past and relationships of other people and places. It is commonplace believed that no one, especially a man, should die outside of his ancestral lands (Moore 1970: 326). Thus one's entire existence is intertwined with specific places and the performance of rituals. Rituals assist in the negotiation of the life crises (van Gennep 1965). The mythical narratives common throughout *Uchagga* dictate that the first ancestor planted his walking staff in the ground and it grew into the tree of the local patriclan. Members of the chiefly lineage were buried underneath the tree. The various ritual activities that require the planting of *msale* re-enact this element of the ancestral past (Hasu 1999: 500).

In the precolonial period the petty chiefs depended on the elders to maintain their power base. The instability of this period has been well documented (see Dundas 1968: 285). It has been noted how chiefly justice was administered by public opinion in lawn assemblies (for a superb review of precolonial and postcolonial customary law in Kilimanjaro see Moore 1970). The precolonial chiefdoms were comparatively wealthy and technologically advanced due to the caravan trade of slaves and ivory through the Swahili coast (see Feierman 1990; Johnston 1886). Moreover it is only with the inception of colonialist regulation that patrilineal chiefly rule was secured as chiefdoms were supported with force in the regulation of tax collection (Moore 1970: 330-2). Nonetheless the point has been made that the chiefdoms were still far from secure and stable. Indeed the chiefs always required the support of their subjects to maintain office.

> "What in the pre-German period had been settled by a combination of wars and diplomacy, was now settled by intrigue, diplomacy, manipulation of colonial officials, and sometime by threats and murders" (Moore 1970: 332)

The importance of the elders within their communities and patriclans remains – something that perhaps contrasts positively against the youth-obsessed beautification culture of the contemporary West. The elders are responsible for much of the education and learning of the young (Marealle 1965; see also Mosha 2000). Indeed the value of the knowledge gained from the grandparents is delightfully related by the proverb '*mafundo ga mku ni matetera*' which translates as 'the teachings of the ancestors are good guides to life' (Marealle 1965: 56). Much of this learning relates to storytelling. Storytelling and linguistic skills are considered art forms in Wachagga culture and some storytellers are believed to be able to intoxicate listeners (Mosha 2000: 50). Through storytelling and proverbial narratives the elders educate the youth. It has been noted that proverbs are in many ways indigenous sacred literature (2000: 56-7).

A recently published study on the Wachagga educational system provides solid insight into the spiritual and symbolic dimensions of their worldly perception (see Mosha 2000).

> "To know in the Chagga worldview means to both have the intellectual information about what is known and to be spiritually inspired by the inherent transcendent aspects of the same known subject...A person not only *knows* that a certain herb has a specific medicinal value, but also *feels* connected to it in the circle of life, and is *awed* by its essential role in the universal dance of life" (Mosha 2000: 30, emphasis in original)

Mosha (2000: 16) reminds the reader that the Wachagga have a holistic approach toward the world with spirituality, intellect, knowledge, emotion, culture and life and death being integratively experienced. Wachagga spirituality permeates every facet of the indigenous worldview. The elders are responsible for the holistic education of the individual. This education system is called *ipvunda* and includes an education for life as well as an education for living (2000: 16). This process continues throughout life and one is awakened to insights and dimensions to knowledge constantly (2000: 17). Some of the education for living concerns respecting the surrounding environment and recognising the sacredness of the landscape (2000: 77). The *mpvunde*, or learning one, is taught reverence for other things as well. Mosha (2000: 89-96) recognises first-, second-, and third-order reverence. First-order or supreme reverence is reserved for the sacred. Second-order reverence relates to the ancestors. Third-order reverence concerns the elders. This third-order reverence is often demonstrated through children being named after certain individuals. For instance, if one is baptised it is customary to adopt the first-name of a biblical or religious figure. In these circumstances the intergenerational name becomes their middle-name. This is again suggestive of a missionary program of accommodation rather than assimilation.

The young show respect by listening to an elder's experience and guidance. Moreover there are linguistic rituals of address and communication that the young and elders must adopt in their discourse (Mosha 2000: 93). This regulation and hierarchisation of reverence reinforces all categories, levels and orders. Thus supreme reverence is articulated and affirmed whenever dialogue takes place. Self-control is taught as part of the *ipvunda* process. It is interesting to note that an important defining characteristic of an elder is the ability to control her/his tongue. Indeed the proverb 'the fools are speaking, the wise are listening' relates to this wisdom (2000: 105). Wachagga philosophy notes that silence, contemplation and reflection are the keys to enlightenment. Thus one must control the urge to talk to capture the essence of an experience. These insights that relate silence and enlightenment are exhibited in a multitude of myths. *Ruwa* is frequently referred to as the 'Greatest Listener' (Mosha 2000: 90; MC/As2/July 2004). Moreover many interviewees professed the belief that talking and noise-making on certain parts of the mountain were likely to have life-endangering consequences such as drowning from excessive rainfall and being lost due to dense cloud impacting upon visibility (e.g. MH/Mh4/August 2004).

History and memory are attractive and powerful characteristics in the Wachagga worldview. Hence elders are revered for they have their own history to depend upon but also have access to the ancestral patriclan memories and histories. When an elder dies, provided the appropriate ritual performance take place, they join the ancestors in the spirit world. These newly deceased ancestors are in closer proximity to the living than the more powerful remote ancestral spirits. There is a common belief that the spirits of the recently dead remain near the location of death whereas the more remote ancestors and the collective embodiment of the patriclan's past are everywhere. A *mwanga* or traditional medical practitioner at Maharo explained "a spirit rests where the body dies if it remains far away from its home it must be brought back to balance the land of the relatives" (MH/Mh48/September 2004). There are various rituals that can be performed to relocate a 'misplaced' spirit. The best method is to travel to the place where the spirit lingers and enact certain ceremonial practices. Obviously this is not always practicable and other devices are used such as orientational rituals, usually involving rocks and soil, which cause places to move. One such ritual was described during an interview with a local *mwanga* who was Mchagga but was taught the arts of traditional medicine by the Pare. Essentially to relocate or translocate a spirit a relative must travel in the direction of the place of death for no less than two days on foot. They must then remember and visualise the deceased whilst obtaining a stone and a small amount of soil. These items need to be returned home. On returning the traditional medical practitioner will perform certain rituals with the soil. The rock should be placed outside the deceased's hut. Further the person who travelled to obtain the items is required to perform a slaughter and leave offerings of meat and local brew to the ancestors so that the incoming spirit will be accepted into the patriclan ancestry. This process relocates the spirit (MH/Mh57/September 2004). The involvement of place and landscape in these rituals is highly apparent. The rock becomes a this-worldly vehicle of movement that provides physicality for the spirit to occupy. It is clear that the ancestral spirits permeate the entire memoryscape and their relation to the landscape is pivotal to the Wachagga religious world.

ANCESTRAL SPIRITS AND SUPERNATURAL AGENCY

Ruwa is sometimes described as the Great Ancestor for he is the origin (Mosha 2000: 8). In addition "Ruwa had his messengers, namely, the departed spirits to whom the living were bade to offer sacrifices at the appropriate times" (Marealle 1952: 61). The Wachagga believe that when an individual dies they exist in the world "in a different form" (Marealle 1965: 58). The physical presence of these spirits is rarely noticed as they are usually encountered in dreams or through the mediation of various animals (1965: 58). The importance of dreams in cosmological beliefs has been explored elsewhere (see Jedrej 1995; Jedrej and Shaw 1992). The non-physical ancestral spirits are capable of consuming libations and offerings and interacting with the physical world. Ancestors are revered. Stories of the ancestors are recited and memories are retrieved and articulated with the highest respect (Mosha 2000: 90). Ancestors are not worshiped as deity figures but are respected as supernatural forces and as media that can communicate with *Ruwa*. The ancestors are believed to be able to reward or punish their descendents through influencing the provision of rain and children (Omari 1991: 179). Respect of ancestors is articulated in various ways. Prayer can be either offered through the ancestors to *Ruwa* or they can be relayed directly to the ancestral spirits themselves. Moreover offerings and libations usually accompany such prayer (Mosha 2000: 91). Reverence is given to ancestors through the education of the young who know that the living must "remember our ancestors, appreciate what they have been and will continue to be, and give them the reverence they deserve as part of our family and part of the human family" (2000: 91). In a similar fashion to that described for elders, respect is shown to the ancestors by naming children after those that remain identifiable in living memory. Thus in such naming customs the past is recast and rearticulated in the present. This is also part of the process of appreciating and remembering.

> "[R]especting our ancestors and respecting everyone and everything makes a person more human, it contributes positively to one's humanization process. It deepens one's education and one's humanity" (Mosha 2000: 92)

The belief in the ancestral *warumu* was particularly evident throughout fieldwork. The present is inherited from the past. Thus in the absence of ancestors the world becomes nonsensical. As one interviewee illuminated "if it were not for my grandparents how could I exist – of course *warumu* are all around and help me with my life" (MH/Mh18/August 2004). The present could not exist devoid of the influence of historic agency. Thus the past influences life everyday in everyway. The belief in the ancestors is absolute as the following two sets of remarks illustrate:

> "I believe in the ancestors, remember them, pray to them, and ask them for the things that I need. They guide me and they protect me" (MH/Mk41/September 2004)

> "I believe in the ancestors and their involvement in everything very much. I do everything I can to appease them so that I might live longer. You should speak to the church leaders they will explain to you that God says 'man must honour and obey his parents'. I must honour my ancestors because they are all my parents" (MH/Mh4/August 2004)

The last quote is most interesting as it demonstrates cultural contextualisation of biblical and religious detail. The interviewee, who was Catholic, may have made

cultural sense of the 4th commandment in the Augustinian sequence 'Honour thy mother and father' (Exodus 34: 28; Deuteronomy 10: 4). It is also interesting that in the Hebrew tradition the commandment of equivalency, which is listed 5th, reads 'honour they father and mother, in order that thy days may be prolonged upon the land which the Lord thy G-d giveth thee' (Orthodox sequence of Exodus 20: 3). Perhaps this further illuminates the teachings that the informant may have received.

Most individuals believe that the ancestors are omnipresent hence they are everywhere. Omnipresence and omnipotence are indicative of the sacred in the traditional cosmological world and in the pervasive missionary teachings. *Ruwa* is omnipresent and omnipotent. The mountain and the sun, traditionally believed to be sacred manifestations of *Ruwa*, are also omnipresent and omnipotent. Hence the environment subsumes divine power and shares sacred potency. It is also believed that ancestral spirits are omnipresent and omnipotent. Numerous informants recounted personal narratives and memories of episodes from their past which they consider to be times when they have been aware of the presence of the *warumu*. The most common of these were recollections of hearing the ancestors singing in the forest. In most cases the singing was investigated. Yet it was impossible to find the source of the songs. Indeed the singing was from all around.

> "The ancestors began to sing their beautiful songs and you cannot help but be hypnotised by them. You want to seek them out to get closer to the voices. You cannot. First they are up-high above you in the forest and then they are far below... It is impossible to find them – if you hear them you should just respect and listen" (MH/Mh1/August 2004)

To the patriclans on the mountain slopes Christianity is a relatively new phenomenon brought by the missionaries and ordained by *Ruwa*. All changes are viewed as being choreographed by *Ruwa*'s will. Therefore everything prior to the missionary arrivals was also choreographed by *Ruwa* and thus unquestionably right. Traditional beliefs are seen as being spiritually ordained and appropriate for the past and as such their ancestral spirits maintain their potency. In the Machame area of the study there was general concurrence that ancestors are not in Heaven as conventional biblical readings might indicate. It is proposed that because they cannot have believed in the concept of salvation through Jesus Christ, because they lived prior to his existence, they obviously cannot have entered Heaven (MC/Nr32/July 2004; MC/Ns18/July 2004). Therefore the ancestors "must exist somewhere for they cannot no longer exist" (MC/Fo5/July 2004). The 'ancestor cult' maintains some validity in the indigenous intelligible world. Normative readings of the bible would proclaim that past people, as well as those of the future, are incorporated in the salvation concept. Indeed Judeo-Christian and Islamic religious traditions concur that all people share the burden of Original Sin. Hence in the Christian gospels Jesus Christ, the sacrificial Lamb of God, suffered the crucifixion to atone for the sins of all people. Perhaps the salvation concept fails to integrate itself within the ancestral potency of the landscape. It is also proposed that the Wachagga understanding of the past, which is linked to biblical events, is not directly correlative with the Western Christian calendar. Indigenes reason that not only was salvation through Jesus Christ only attainable after the time of his existence but also that it was only possible after the arrival of the "missionaries who brought the new words of *Ruwa* with them" (MC/Ud18/July 2004). Therefore the biblical appreciation of the Wachagga telescopes two millennia into less than one and a half centuries.

HUMAN REMAINS AND SHRINES TO THE ANCESTORS

The Wachagga are intrinsically attached to the landscape through multiple cultural practices, beliefs and comportments. Human remains, especially skulls, have traditionally been regarded as sacred by the Wachagga.

> "The Chagga people regard the skulls of their ancestors as sacred relics, particularly those of chiefs which are greatly revered. In Marangu I recently asked to see Mareali's skull and though the chief offered to show it, I refrained from availing myself of the offer because it was obvious that the elders were nervous about it probably fearing that I might appropriate the skull" (TNA/SE/AB988A)

Due to the sacredness of the skulls and other bones of the ancestral elders and chiefs and their situation within inherited patriclan lands the land is imbued and permeated with religious meaning. This is further sedimented through ritual activity involving the bones and various sites of ritual action. The ancestors are tied to the procreative potential of the living. Sterility is a potent negative force that can potentially infect the earth (Moore 1976: 358). Reproduction is immortality – a perpetual bond between ancestors and descendents. Spirits can only exist through the memories and offerings of the living. In the absence of these rituals of remembrance and libation ancestral spirits become dissatisfied and eventually perish (1976: 358).

> "The man or woman who has no offspring dies, and his line dies, and that is the end of him, or her, the complete, absolute end. It is not just the end of life in this world, but the total end" (Moore 1976: 358)

Thus ancestral sites and shrines are associated with both life and death and their potency is further enhanced by the presence of human remains. The ancestors are revered through special shrines dedicated to them (see **figure 14.1 and 14.2**). These shrines memorialise them through processes of re-remembering, co-remembering and de-remembering. These shrines called *mbuoni* are sacred places surrounded by *msale* trees. Access to these shrines was traditionally restricted to adults (Mosha 2000: 92). It has been implied that these shrines have more recently been replaced with graveyards (2000: 92). This is true on a functional level as the rearticulation of human bones is seldom practiced anymore but on a symbolic level these shrines remain vital components of the religious landscape. Graveyards, or graveplots as they are better described, have indeed become respected places in their own right. This is linked to the intrinsic cultural sacredness of human remains and ancestral spirits. Hence both graveplot and *mbuoni* are important religious characteristics of the landscape.

The *mbuoni* are traditionally potent sites exuding sacred power. They are highly respected areas. If one goes there one must be respectful and perhaps take offerings of meat, local brew or milk (**figure 15**). These are the sites where traditionally exhumed bones, particularly the skull and humerus, were moved to and rearticulated within. These sites see considerable ritual action even in the contemporary period including prayer, the deposit of libations of slaughter meat, and the pouring of local brew and/or milk over the site. *Mbuoni* are always gender-discriminate. Moreover so are the rituals associated with them. For instance, local brew must be poured over a male *mbuoni* and milk over a female one (MR/Ly19/August 2004; MH/Mk5/August 2004). *Mbuoni* may contain the bones of single or multiple dead. Some areas have many singular *mbuoni* in close proximity whilst others have collective *mbuoni* containing the rearticulated bones of many individuals. The commoner collective *mbuoni* are also gender specific.

The Lutheran church near the area of Marangu where some of this research was undertaken has fairly successfully eradicated this rearticulation practice (MR/As27/July 2004). Nonetheless there are still some families and patriclans who continue this traditional action. Much to the displeasure of church elders it has also been known for slaughters to occur at the sites of graves as an alternative to exhumation (MR/As27/July 2004). The *mbuoni* does not only witness ritualised rearticulation or libation activities following an individual's death. The main function of the *mbuoni* is to serve as a communicative site facilitating dialogue with the proximal ancestral spirits. The commonest method of conversing with the ancestors is through prayer associated with offerings. Essentially heads of livestock are slaughtered and meat offerings are placed on banana thatch or *msale* leaf and deposited on the *mbuoni* where they are left unattended overnight. If the meat has disappeared the following morning it is thought the appropriate ancestral spirits have consumed the offering and their accompanying prayers will be answered. Typically such offerings are made at the *mbuoni* when individuals are sick, when rain is needed, when individuals want babies and so forth (MH/Mk8/August 2004). These shrines facilitate the mediation of the past with the present, the dead and the living, and *Ruwa* and the patriclan. It is a place of respect and power.

Figure 14.1. Photograph of male mbuoni from Marangu (note human humerus at base of shrine)

Figure 14.2. Photograph of female mbuoni from Marangu

Figure 15. Photograph of libation pouring instrument discarded at mbuoni

RITUALITY: SLAUGHTER AND THE LANDSCAPE

There is much literature that considers the pan-African practice of animal slaughter and sacrifice (see Bierlich 1999; Cochetti 1995; Dalfovo 1997; Hojbjerg 2002; Rasmussen 2002; Walsh 1996). As in many of these studies the ritual slaughters performed by the Wachagga serve to amplify and exaggerate attachment to the land. The term slaughter is utilised in preference to sacrifice because the ritual is usually conducted to mediate between the living and the recently deceased ancestors. Moreover the slaughters performed in worship of *Ruwa* or the remoter ancestors might be classifiable as sacrifices but the term is used cautiously owing to it carrying much unwanted cultural baggage.

Historically Dundas (1968: 136) noted that the sacrifices he witnessed were "singularly unceremonious and destitute of the ritual one would expect". In fact those reported elsewhere (e.g. Hasu 1999) and observed during the recent fieldwork are actually highly ritualised affairs. The ritualisation does not adhere to Western cultural expectations of religiosity but it is apparent and it is far from understated. Fuller appreciation derives from a more comprehensive understanding of the culture and historical religious system. It has been recorded that entrails and intestinal tissue of the sacrificial animal were often

examined by the elders for divination and prophetic purposes (Dundas 1968: 142). It is still the case that intestines are kept until the day following the sacrifice for consumption and divination (Hasu 1999: 517). Dundas (1968: 150) also recorded that sacrificial rituals performed for the ancestral line of the chief required *kisuku* animals – animals that were entirely black, white or red. In the past these public sacrifices were required for ruling chiefly lineages as well as those that previously had claim to the land (1968: 151). Moreover in the past every patriclan had its own sacrificial customs (Marealle 1965: 58). According to the literature sacrifices were offered in specific places where there was a history and tradition of such ritual events – such as in rivers, lakes or under big trees (Marealle 1965: 58). It is customary to offer the stomach contents and intestines of sacrificial animals to the ancestors (Moore 1976: 366). This practice continues and again highlights the life-bringing symbolic dimension to faeces. The digestive-intestinal contents are offered to the ancestors so that they may be nourished and so that they can interject on behalf of the living into the affairs of *Ruwa* (see Moore 1976). Furthermore the mountain and the ancestors feature in the performance of ritual slaughters intended to accelerate the recuperation and convalescence of the ailing. The following was recorded recently in the Vunjo district and is part of a prayer recited by a group who were sacrificing a bull whilst facing Kibo:

"We know you, Ruwa, Chief, Preserver. One who united the bush and the plain. You, Ruwa, Chief, the elephant indeed. You who burst forth people that they lived. We praise you, and pray to you, and fall before you" (Mosha 2000: 8)

The most common sacrifices are performed as a component of mortuary rituals. This sacrifice is a transformative medium of transportation for the deceased. The sacrificial animal is representative of the reproductive character of the deceased for example a gelded goat would be slaughtered for an old man or a ram for a middle-aged man (Hasu 1999: 470). The Wachagga do not believe that sacrifice and ancestral beliefs are contradictory (1999: 487). Indeed sacrificial action is a moral prerequisite of society. Sacrifices (re)centre people within the moral community. The sacrifice expresses the social order. The sacrificial animal is a metaphor of sociality that requires disintegration and rearticulation – transformation into a new physical and moral state of being. Thus substances become ritually important because of their various synergic states of congregation and association (Werbner 1989: 115-7). Death and consumption incorporate the offering with its recipients encompassing both the ancestors and the living.

Beliefs and rituals are not uniform or constant but inhabit the arena of negotiation and contestation. It is seen in *Uchagga* that sometimes *msale* leaves are tied around the neck of the sacrificial animal. Sometimes the sacrificial animal must be brought out from the home of the deceased when it is moved to the slaughtering site in the *kihamba*. The movement is thus analogous to the historic transfer of bones from the primary to the secondary burial site (Hasu 1999: 503). The greater the length of time taken for the sacrificial animal to die the better impact can be anticipated. The sacrificial animal is a metaphor for the body of the deceased. The sacrifice is involved in the processes of unmaking. The congregation and distribution of the sacrificial meat is highly ritualised (see 1999: 496). The consumption and ingestion of sacrificial meat disaggregates the person and incorporates them within the bodies of the living and the form of the ancestral dead. Individuality is transformed into the social/collective whole. Moreover parcels of meat are taken to graves and *mbuoni* as offerings (1999: 497). The sacrifice is thus a performative act that articulates the social order by embodying all categories of people. Thus the ritual reaffirms relations and identities. Hasu (1999: 507) argues, in a Durkheimian fashion, that the sacrifice represents an externalised social whole that becomes incorporated with the ancestors.

Following the initial sacrifice that accompanies a Wachagga burial further sacrifices need to be performed. The most important of these is performed after forty days. This second slaughter was historically performed after the period of mourning elapsed but the Islamic practice of *siku arobaini* has accounted for the delay (Hasu 1999: 495). A comprehensive description of funerary sacrifices has been provided elsewhere and so will not be repeated here (see Hasu 1999: 495-518) rather what requires underscoring are cultural understandings and beliefs relating sacrifice and the environment. The performance of ritual slaughters, including funerary sacrifices, can potentially instigate positive repercussions. For instance, they can evoke blessings from the ancestors and *Ruwa*, cure the sick, bring rain and amplify harvests, promote the fertility of animals and womenfolk, and integrate one's children within the family if some were born outside of wedlock (MR/Mb2/August 2004; MH/Mh6/August 2004). Slaughters are anything but the mere profane means of killing cattle prior to consumption. Certain animals make better candidates for slaughter. Whilst in the field the slaughter of cows, goats and to a lesser degree chickens were regularly noted through interview and participant observation – although there were some rarer indications that ducks, pigs and sheep were sometimes slaughtered (see MH/Mh18/August 2004). First-born animals or those coloured black and white are preferential and this relates to the historical ordination of such practice.

It should be mentioned that some Wachagga do not actually perform slaughters themselves but instead purchase the required pieces of meat from a butcher so as to conduct rituals. This suggests a dilution of the rituality but may actually represent modification based on performative change. The Roman Catholic institutions in Maharo have exhibited a preference for these adapted activities although clearly they refrain from positively encouraging any such episodes (MH/Mh19/August 2004). The Wachagga believe that the blood from the slaughtering enriches and nourishes the ground and makes the area particularly spiritually potent. In some ritual slaughters blood is collected and subsequently poured into the ground whilst prayers are articulated (MH/Mh19/August 2004). In others blood is permitted to drain directly onto the ground after the throat is severed. The site of the slaughter thus absorbs the life-giving properties of the blood. Other ritual slaughters require the utilisation of blood elsewhere and thus minimal amounts are permitted to drain into the ground. It is conceivable that this may be a recent ritual development, instigated by diminishing resources, that better utilises the nutritional value of blood.

The church elders in the Roman Catholic areas promote smothering and suffocation as optimal methods for killing cattle prior to consumption. Nevertheless most slaughters are conducted with a knife (MR/Ms32/August 2004). Numerous interviewees confirmed that 'covert' slaughtering and offering did still occur, particularly at night, and when they did they were mostly during times of social and economic stress (MH/Mk31/August 2004). It is, of course, possible that the covertness of the slaughters is overstressed and that their performance is a community-wide open secret. Slaughtering is within the male sphere of control and women are excluded from participating. Even a local clergyman reported his involvement in these ritual slaughters although he did quantify his participation by noting such activities "connect us with our God through our elders and the mountain" (MR/Ms9/August 2004). Others are far more explicit with their faith and reliance in the necessity of performing ritualised slaughters.

> "Let not the church leaders cheat you into thinking otherwise, even up to today we are making sacrifices of goats and cows, without these we would not survive" (MR/Ms18/August 2004).

Slaughters are usually performed at specific places which are differently characterised and described throughout the various regions of *Uchagga*. These characters will be highlighted through the *kiungu* and *kiriau* of Maharo and the *ndalla / shenyoni* and *ifumvu la mkuu* of Marangu. Slaughter sites are particularly informative illustrations of the processes involved in the sedimentation and inscription of collective and individual memory upon the landscape as they have entire patriclan and family cycles played out upon them through the slaughtering that accompanies culturally significant events.

The *kiungu* are comparable to *ndalla* or *shenyoni*. They usually consist of two or three large trees between which slaughters are conducted. In the area neighbouring Maharo smaller *kiungu* with one tree are known as *kiriau*. These places are the sites utilised in the performance of routine slaughtering. It should be noted though that slaughtering is never deemed an utterly routine affair. Slaughtering occurs at special occasions and ceremonies but at the same time a slaughter makes an occasion (MH/Mk8/August 2004). If somebody decided to slaughter a head of cattle it would be a planned affair. The notion of a spontaneous slaughtering caused confusion – it is something to be looked forward to and anticipated (see MH/Mh59/September 2004). Most families and patriclans will have their own place for routine slaughter. Ordinarily this will be in close proximity to the domesticated space of the *kihamba* buildings. These sites are intrinsic to funerary rites. Ritual activity after death is a highly elaborate affair and ordinarily involves the slaughter of head/s of cattle. These slaughters are essential in maintaining the cosmological order for without them the deceased spirit cannot become united with the former elders. This unification process is highlighted in the funerary tradition that the father or eldest surviving son must, to accompany the slaughter for a dead person, proclaim "and now you can become a member of the dead clan" (MH/Mh14/August 2004). Slaughtering at these sites is usually not directed to *Ruwa* but rather toward the ancestors in solicitation of their prayers and influence. Thus the ancestors are the intermediaries between *Ruwa* and the living. The slaughter as ritual action serves to articulate and cast the relations between the living and the dead. These slaughter sites embody this relationship and these articulations through repetitive action and the very act of facilitation. The underlying theme of unification is also apparent in that slaughtering unites one with *Ruwa* through the ancestors. The past and the present and the dead and the living are related, connected and ultimately united through the performance of slaughters. During the funerary ceremonies there will be subsequent slaughters to memorialise and (re)call specific individuals who have died and the patriclan collective dead.

There are far fewer slaughter sites nowadays than there were historically because of the prohibitive influence of the world religions. Moreover many of the smaller ones are being abandoned (MH/Mh46/September 2004). Nevertheless all individuals can tell you the place of their patriclan or familial slaughter site. Furthermore everybody continues to be associated with an active site. Whilst preserved in living memory these sites retain their sacredness. This is another example of the religious dimension of landscape memory and the sedimentation of the past.

> "We have abandoned our clan kiungu and now we go to the big one in the next village [Maharo]. It has been destroyed by the felling of trees and crop growing. The members of my clan and I do sometimes put a few pieces of meat at the site" (MH/Mk36/August 2004)

The *ifumvu la mkuu* is another form of historical slaughter site. It has been noted that *fumvu lya mku* or the 'mountain of ages' is a term that is used to glorify *Ruwa* (Marealle 1965: 57). The term is an expression for *Ruwa* because in the Wachagga cosmological worldview to last eternally, and hence possessing both histories and futures, is divine. The patriclans of the Marangu area also refer to a specific place as *ifumvu la mkuu* this relates the particular place to the divine but also designates the site as a medium for communicative prayer. Moreover *Ruwa* is thought to dwell at these sites (MR/Ms37/August 2004). Traditionally these places were the location of patriclan worship, slaughter, and circumcision rites. The use of these sites has diminished as prayer – especially that which is accompanied with ritual slaughter – is no longer performed overtly due to Christianity and the religious apathy of the younger generations (MR/As21/July 2004). This is not to imply that these places no longer hold significance indeed on the contrary they are powerful and important components in the local environment.

During the recent fieldwork there was occasion and opportunity to explore numerous sites referred to as *ifumvu la mkuu* and all of them were overgrown with dense vegetation and difficult to access. This perhaps implies that these places were uncommonly visited. It should be remembered though that sacred sites

Figure 16. Photograph of Shao patriclan ifumvu la mkuu

throughout East and West Africa are frequently overgrown with vegetation but that does not necessarily correlate with infrequent use (Insoll 2004: personal correspondence). If they were being visited as informants suggested (see MR/Ms2/August 2004) then access and penetration would be of considerable difficulty. It was originally suspected that interviewee's comments concerning these sites had been overstated but it later became apparent that these sites were being left dormant and their fertile potential left unexploited. With land in such short supply and harvests somewhat erratic these sites must possess some latent importance and mythic prominence. Usually the land had an 'owner' but in cases where the owner was of Wachagga origin there was no development or interference. For instance the Shao patriclan *ifumvu la mkuu*, where in the past the Mamba Chief Lemnge performed many rituals, belongs to the Agape Secondary School. This site is deemed particularly important as it sits on the intersection of three wards – Mamba, Marangu and Mwika – and is thus considered very special. Cohesion and unity are of considerable cosmological significance. The site remains undeveloped and unexploited (see **figures 16 and 17**).

Elder's Hill (-3°10' 37°45) is another example of *ifumvu la mkuu*. This shrine on the land marks the location of the ancestral hunters and the arrival and settlement of the 'clan of the chiefs' high up above the forest on the mountain (Tagseth 2003: 22). Again the relation between historicity and divinity is apparent. Elder's Hill is near Mandara Hut (formerly Bismark's Hut) inside the National Park boundary. The hillock is wooded and neighbours a small water pool. Ritual slaughters and offerings are performed at Elder's Hill particularly during time of economic stress (MR/Ms33/August 2004).

> "Chaggas do practice traditional customs and hold beliefs in the old ways because they know of and have received the blessings of Ruwa. Christianity has changed our binding with Ruwa, but we still have to depend on the older ways to get what we need" (MR/Ms14/August 2004)

Interviewees reported that "Elders Hill is where our traditional *Ruwa* is" (MR/Ms29/August 2004) which suggests that perhaps a duality exists between the

60

traditional and Christian *Ruwa*. This would be erroneous for it is commonly conceptualised that *Ruwa* is divisible into essential elements of supernatural force and potency. *Ruwa* can occupy any form and is the relational instigator of everything including the past and the present. The traditional aspects of *Ruwa* are seen as being more approachable and communicable through the environment than the newer ones. Thus the traditional manifestations of *Ruwa* deserve to be attended to and/or exploited. It is also interesting that when informants recall conducting rituals and slaughters at Elder's Hill the

expression 'go to the mountain' is frequently used (MR/As18/July 2004; MR/Ms29/August 2004). It should be remembered that the *ifumvu la mkuu* is not directly venerated – it is special and powerful in itself only because of its real power as derived from it being componential to the mountain. Similarly *Ruwa* is ultimately sacred but other manifestations of the supernatural are incorporated within the numinality inherent in places and things – water pools, slaughter sites, caves and so forth.

Figure 17. Map depicting the location of the Shao patriclan ifumvu la mkuu and the various political/administrative units incorporated in the 'dialectic-soil zones'

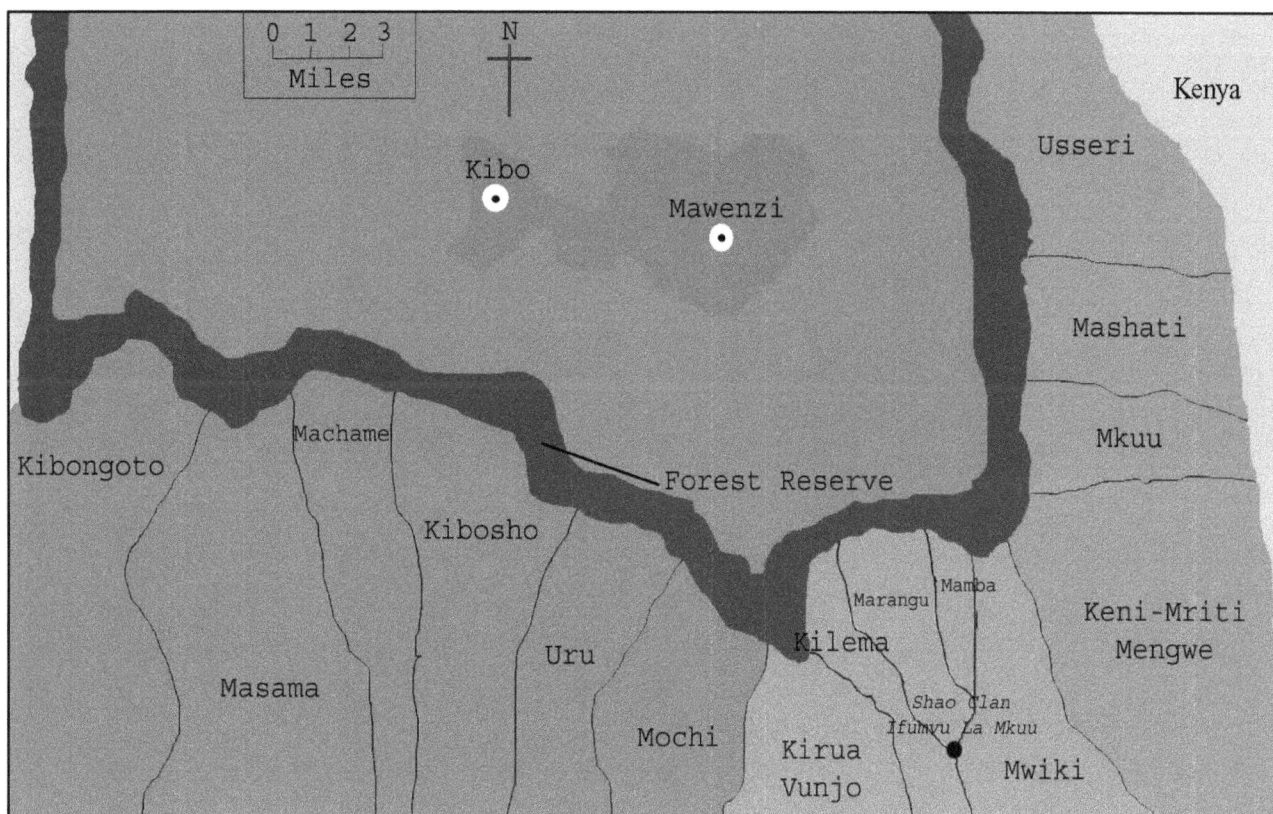

Nomenclature:

Kimashami dialectic-soil zone   Kivunjo-Kimarangu dialectic-soil zone

Kihorombo dialectic-soil zone

CHAPTER EIGHT

INSCRIBING THE LANDSCAPE: LANGUAGE, PLACE AND OWNERSHIP

PLACES AND LANGUAGE

Place names are another important factor in the sedimentation of landscape, memory and attachment to the land. Places are invariably orientated-toward through local cues concerning patriclans and individuals both alive and deceased. The past is constantly reinvented in the present through naming processes. The perpetuation of the old order in the present recasts and reiterates memories and histories (Darvill 1999: 107). The present re-remembers past events, action or circumstances. The landscape is named and located and consequently space and time are fused (Tilley 1994: 33). Thus it is no surprise that the place-name Marangu means "much water" (Bailey 1968: 163). Myths and legends enhance this process and have been explored in some depth elsewhere (see Schama 1996). The past is made explicit through the reference and humanisation of place. Indeed place is by definition humanised space (Tilley 1994: 14). In the Wachagga cosmological worldview all land is (super)humanised through the coordinating projects of the past and the future. This is partly because all land, in being fertile, is the farming potential of somebody but also there is something more subtle apparent. Land is empty without the humanising processes of occupancy, agency and association. The response to the question 'who owns the Ngira land?' clarifies this multi-layering of history and meaning in a local context.

> "Ngira is Nkya's old place. It has many owners. We all go there to graze goats and cows and sometimes to see Swai and Minja who live nearby. Ngira is a special place so God or the other villagers will punish you, if you damage it… It is not my place but everyone's place. Some people actually live there but we all share in it" (MH/Mk6/August 2004)

Nkya was one of the brothers who first settled in the Machame area (MC/Fo4/July 2004). Swai and Minja are patriclan names but the respondent was referring to two specific individuals. The importance of the historic agency associated with the place necessitates a belief in some form of punishment ensuing from acts of 'sacrilege and desecration'. The notion of punishment is interesting, in this context, for villagers have limited power to 'punish' in any formal authoritative sense. Nonetheless they can ostracise and marginalise individuals. They may also petition the village council, subcommittees, or other related organisations to take punitive measures if necessary. Even in the contemporary era these types of social infringements will be resolved through the judgement of the local chief (MR/Ms9/August 2004). It should be highlighted that social isolation in a society with very low mobility and a high sense of communality would offer considerable deterrent.

There is of course the Wachagga tribal-ethnic identity acting as homogenising glue throughout *Uchagga* which characterises the land. This tribal-ethnic identity situates, fixates and relates the Wachagga to the landscape. Despite complex tribal-ethnic origins (Marealle 1952: 58) and the presence of neighbouring tribes situated in similar environmental niches (e.g the Arusha and Meru; see Spear 1997) the Wachagga have a distinctive identity. Difference in the relations between the patriclan and the landscape is a dominant factor in such processes of identification. Settlement place names frequently relate to the earliest settlers and so exhibit these tribal-ethnic affiliations. Moreover the local dialects of the Wachagga have embodied and assimilated much from these tribal-ethnic neighbours. As already mentioned the Wachagga have traditionally been documented to be an absorbing people (see Stahl 1964).

There are many local mutually unintelligible dialects of Kichagga (Tagseth 2003: 168). The principle dialectic versions are Kimashami, Kivunjo-Kimarangu, and Kihorombo (see Hinnebusch and Nurse 1981; Nurse 1979; Nurse and Philippson 1980). These dialects are not understandable outside their respective geographical boundaries (MR/Mb16/August 2004). Languages and cultures fix local people to the land. Thus the land of the Kihorombo-speakers is a physical as well as a cultural conceptualisation. It is interesting to note that the educated and migratory sections of society tend to conceive of Kichagga as the language of women and/or domestic arrangements. Furthermore these groups believe that business and other 'important' discussion should be in Kiswahili (Hasu 1999: 468). This imposition of culture upon the physical environment is also demonstrated in a recurrent awareness that the culture is grounded in a literal sense. For instance one interviewee commented that "the Pare should learn to speak Kichagga because they come from the same soil" (MH/Mh51/September 2004). It is interesting to note that there appears some correlation between the broad dialectic zones and regional soil characteristics. If soils were to be categorised in terms of the general indigenous dialectic zones transcribed upon them the description would highlight subtle correlative differences. The Kivunjo-Kimarangu soil could be depicted as being a brown humic loam, difficult to till with high organic content. Similarly the Kimashami loam could be described as having a heavy brown composition, high water content and high organic content. The Kihorombo soil would essentially be described as reddish-brown, dry and susceptible to high erodibility, and as having less of an organic constitution. These dialectic-soil zones are depicted in **figure 17** (above).

The local conceptualisation of dialectic differences also relates the environment and the sacred. It also highlights correspondence between landscape and the syncretistic religiosity of Christian missionary teachings and more traditional comportments. The story of the Tower of Babel (Genesis 11: 1-9) describes how close biblical man came to reaching the divinity of the Old Testament – manifestly and symbolically – with one language. God punished biblical man by scattering the people throughout the land and, in confounding their tongue, created divisive language barriers so that such enterprises could never be replicated. This biblical narrative was familiar in all the transect areas (e.g. MC/Ud6/July 2004; MH/Mk19/August 2004; MR/Ly20/August 2004). This prevalence and cultural permeation is not simply explainable in terms of the indoctrinability of individuals subjected to institutionalised reinforcement strategies but also relates to the cultural intuitiveness of the content. The permeation of the narrative is due to the integrative theme of unity that has traditionally been associated with sacredness. Through unification and cooperative togetherness biblical man was in a position to attempt the impossible. Moreover the narrative facilitates the metaphoric conceptualisation of the processes of dialectic separation. The environment was choreographed and ordained by *Ruwa* to promote disharmony and precipitate the existence of mutually unintelligible languages/dialects within one tribal identity. The salutary parable of the Tower of Babel religiously attunes the Wachagga to the sacredness found in the landscape.

> "Languages have been decided upon by Ruwa. He wants people separated and the mountain makes the clans grow apart with rivers, forests and languages. This is how things have been decided and how we must live" (MR/As14/July 2004)

RIGHTS OF OWNERSHIP: DWELLINGS AND LAND TENURE

Traditionally hut building throughout most of precolonial sub-Saharan Africa was considered a communal rite that consumed communal materials and therefore resulted in communal property (see TNA/NA/11/677). This cultural understanding was something the British colonial administration went to great lengths to eradicate for they saw it as a massive stumbling-block to the development of ideologies of ownership, self-improvement and commerciality. Thus land, dwellings and property were

of communal as well as patriclanial and familial value. Land, dwellings, property and the means of realising their potential i.e. labour and building resources were controlled and sequestered by the petty chiefs. Although this regulation was probably orchestrated through pervading public opinion (see Moore 1970). To the colonial administrators the result of this customary law was that individual hut-dwellers would have no incentive for developing their residence because it was not actually their possession. If they needed to move to another location they were unable to sell either the dwelling or its construction materials. They would have to petition the chief in order to leave and he would decide its fate – either its demolition or allocation to another individual (TNA/NA/23/SF/1).

The relief of the administration can clearly be observed after the decision of the native courts of the Nyanza Federation (TNA/NCNF/1930/185) that permitted the removal and transportation of building materials, but not vegetation or other land-resources which remained the property of the community, by a dweller in Urima.

> "The hardened cake of custom is already crumbling, and even all the influence of the Council of Chiefs will not avail to prevent the spread of the idea of individual ownership, not only as regards huts but also as regards land" (TNA/NA20239/26)

To a certain degree the indigenous Wachagga conception of ownership is not directly comparable to the Western capitalist version one may be familiar with. There are levels of ownership. Land, for instance is possessed and inherited in a patrilineal fashion. The traditional land allocation granted/ordained by the chief was marked out with *msale*, a sacred variety of local tree, which was planted along the authorised boundaries. The planting of the *msale* was a highly ritualistic act that was witnessed by all concerned. A common belief throughout *Uchagga* is that disturbing or digging-up *msale* induces premature death (e.g. MR/Ms2/August 2004; MH/Mk14/August 2004). Such beliefs were/are effective in curbing boundary manipulation and encroachment. Boundaries and ordained rights to land were/are permanent. To this day *msale* is not uprooted without the local chief's blessing. The planting of *msale* also articulates one's identity as being Wachagga. The ownership of land thus maintains a residual dimension that draws its meaning from the past. Fathers pass parcels of land to their youngest and eldest sons. The present inherits the past. The living inherits the possessions and experiences of the ancestors. Incidentally daughters migrate from the household exogenously and become the concern of another man and patriclan. At the same time land is 'possessed' by particular patriclans. Moreover the former chiefs, who are no longer recognised by the Tanzanian authorities as having a political role, are also accredited with some manner of possession rights. There was reference to mangiates and chiefdoms in interviews conducted along all project transects (MR/As26/July 2004; MC/Ns13/July 2004). It could be argued that this is another example of the past being rearticulated and culturally ritualised in the present through processes of memorialisation.

The former chiefs and mangis are still very much respected authority figures within the local communities they formerly governed. Although they have no state-sanctioned power their traditional positions and customary duties result in their consultation about local affairs. Ward secretaries, administrative officers and village councils are the key political posts/institutions yet despite discouragement from more centralised authority chiefs/mangis and their successors are frequently approached to resolve local issues/concerns. They still very much adopt functional roles in their communities. They are leaders who possess the wisdom of their ancestral positions. The chiefs'/mangis' possession of land imposes on their former chiefdoms/mangiates. Similarly patriclan ownership of land will normally relate to their attachment to the landscape and surroundings in the past.

It is important to note that ownership of land also is also a concern of the supernatural. This does not simply relate to the supernaturality of the ancestors, history and elders but also to the supernatural guardians/owners embodied in the land. These take the form of spirits, usually these are linked to *Ruwa*, which reside in certain features within the landscape especially rocks, water pools and large trees. The spirits embodied in these features factor in the likelihood of such features being used as locations for the performance of ritual action. These considerations inform that to the Wachagga land is not divisible into direct proportional ownership instead individuals and patriclans are bound within overlapping schemes of dynamic proportionality based on notions of cultural attachment, memory, access rights, inheritance, historicity and supernatural agency. Thus the understanding of land as a communal rather than a private resource and site of construction, despite the efforts of the colonialist policies, remains integral to the Wachagga identity. Nonetheless one should be careful of establishing an unsophisticated dichotomisation between colonial individualistic capitalism and Wachagga communality. The pre-colonial Wachagga were hierarchical and it is difficult to believe that with such intricate trading networks and high agricultural productivity elements of ideological capitalism and individuality were not adopted.

CHAPTER NINE

SPECIFIC FEATURES OF SOME MEMORYSCAPES IN UCHAGGA

Wachagga memoryscapes have been considered thematically and elaborated with case study material culled from fieldwork. Consideration, however, needs to be given to the specific as well as the communal dimensions to the mythic worldview in ventures such as these. What follows are descriptions of the locally perceived memoryscapes within three research localities. It would be academic fallaciousness to attempt to describe any memoryscape in its entirety. For as Dundas (1968: 36) notes the landscape is "altogether a repository of mythical beliefs". Thus the material presented herewith does not purport to be exhaustive or comprehensive, but rather what has been included has been done so because of its local importance and its poignancy to the wider discussion. Moreover there will be no repetition of material highlighted in the thematic section.

SIENY CATCHMENT FOREST

In the Machame study area local memoryscape research was focused upon an area high up on the interview transect. In particular examination was concentrated on an area known as Sieny Catchment Forest (-3°10' 37°12'; see **figure 18**) which lies between the villages of Foo, Ngira, Uswaa and Nronga. The area exudes sacredness through its environmental potency. The cultural perspective that unification and integration correlates with the sacred very much informs local perceptions. The area is thought to relate the four neighbouring settlements and is an intercommunion of the history of these (super)humanised places. The villages are separated by a valley gorge that impairs access. The valley floor marks the confluence of the Marire River and the Nanwi River. The entire area is a convergence zone. The visually stunning natural path known as Daraja la Mungu or 'God's Bridge' (see **figure 19.1 and 19.2**) joins the areas of high relief by bridging the Marire River. The bridge unites the local settlements.

As previously noted the origin and historical dimension of things are particularly important forces in the Wachagga cosmological world. Sieny is strongly associated with one village in particular – Ngira. Ngira is of Kimashami dialectic origin and the name conveys meanings of a safety and concealment. The place is associated with one of the first settler of the Machame area detailed in the local oral history – Nkya. Oral traditions note that three brothers were the first to arrive in Machame: Nkya, Mushi and Leama. Nkya settled in the area and is perceived as being the ancestral origin of the Ngira village. Mushi and Leama settled higher up on the mountain. Nkya is thus one of the most remote but important ancestors (MC/Fo11/July 2004). The area is locally believed to be the dwelling place of good and evil power and spirits. It is interesting to note that the stones for the local church Sheny Kwa Kkira 'Come to the Saviour' were taken from the Sieny Catchment area. Informants noted that this action was motivated more out of a desire to harvest the religious potency of the area for the church rather than necessitated through resource availability (MC/Nr29/July 2004). Nonetheless it should be noted that the rock in the area does make good building material and this is apparent in its use in other local projects. Whenever rocks are taken from the area deference is afforded through ritual action. For instance when indigenous patriclans take rocks from Sieny and from the nearby Masama Ngira, for building and grinding-stones, they are obliged to pour milk and/or local brew before they do so as libations. The stones embody these places and retain forces of sacredness and protection. Thus the home built of such stone is blessed and any food processed on such grinding-stones resonate safety and spiritual nourishment (MC/Nr4/Junly 2004).

There are many naturally formed caves in the Sieny area; the larger of which have specific names, ritual functions and associated historical importance. For example, Nkyeku and Kyumbe caves are important due to past ritual practice. It is interesting that both caves are considered in female terms. Nkyeku for instance is the local term meaning 'oldest local woman'. The cave is important to the local people as a historic medium with *Ruwa*. This notion is still very much discernible in the villages although the local church elders try to discourage and undermine these beliefs (MC/Nr2/July 2004). Traditionally the local chief would regularly dispatch a leading member of the Nkya patriclan to the Nkyeku cave with offerings for *Ruwa*. This would usually be a goat which would be slaughtered. If the offering was accepted indictors would appear on the rock face above and within the cave (see **figures 20.1 and 20.2**). This practice has apparently ceased although some villagers do visit the cave occasionally to read the indicators (MC/Nr2/July 2004; MC/Nr19/July 2004). The indicators were/are means for the indigenous patriclans to anticipate, through divine prophecy, future local events. The notion of the past providing access to the future is not exclusively exhibited here. Indeed the notion of ancestral intervention is another common example (see above).

Figure 18. Sketch map of places of power within the Sieny Catchment Forest area

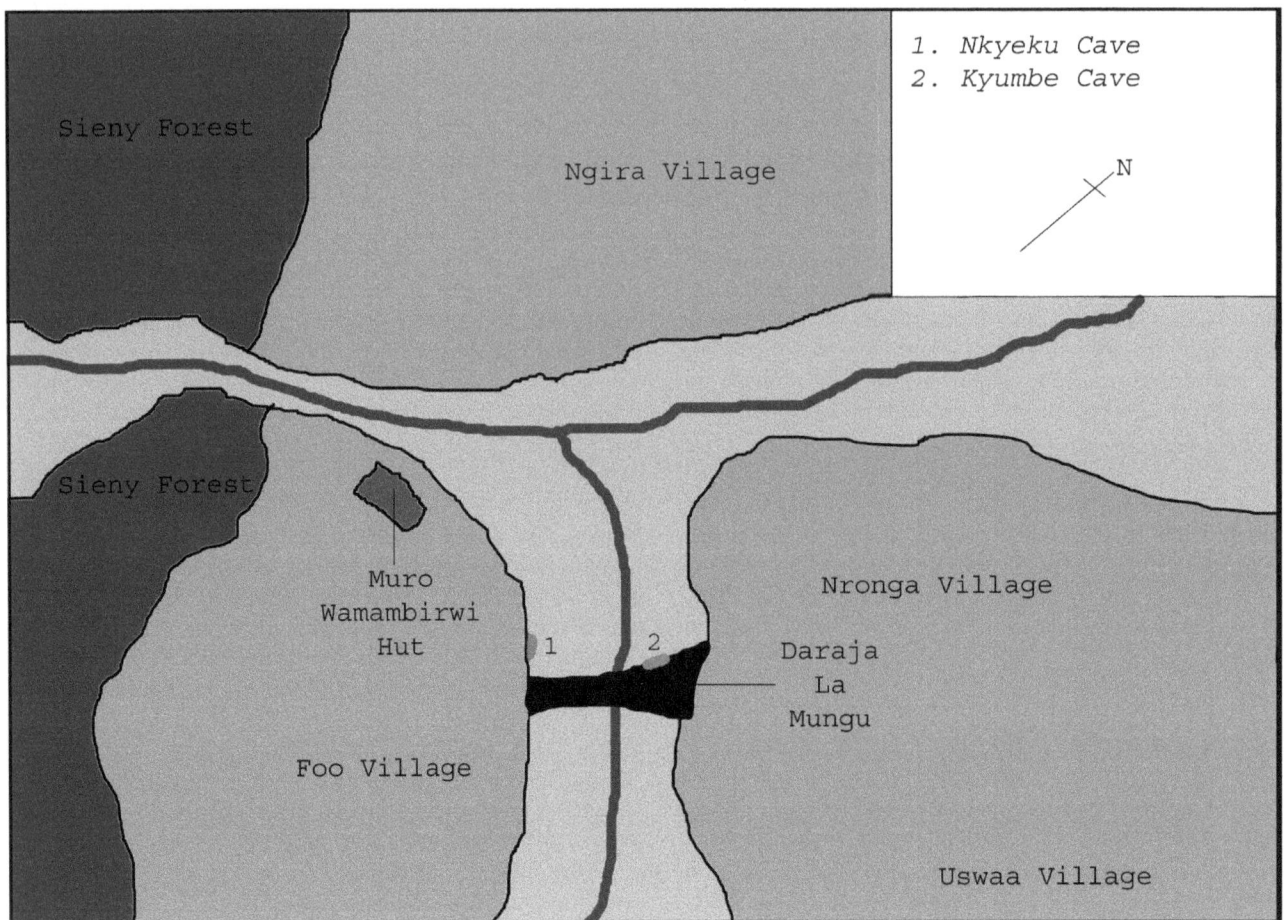

The precolonial Wachagga conceptions of time were cyclical in nature rather than the linear version brought about with modernity. Thus the past was related directly to the future through supernatural connectivity. The future was communicable through the indictors of the cave. The indictors were the colours red, white, green and black. Local interview revealed a consensus that the colour white is associated with good harvest and peaceful times ahead and the colour red with famine and war (MC/Nr14/July 2004). Black was usually associated with disease although some locals expressed a belief that it symbolised the requirement of further offerings (MC/Nr2/July 2004). There was little consensus over what green indicated – some thought rain (e.g. MC/Nr2/July 2004) and others that a local elder was going to die (e.g. MC/Fo34/July 2004). Within 50m of the Nkyeku cave sits the Kyumbe cave and the two are frequently described as being related with the notion of 'sisters' being commonly articulated (e.g. MC/Nr15/July 2004). The Kyumbe cave is referred to in English as the 'barren cave' owing to its traditional use as a ritual means of supernaturally increasing fertility in barren women through ritual performance. Interviews suggest this practice is still widespread. Barren women are taken to the cave by a patriclan elder where she will stay for a length of time dictated by an authority figure – traditionally a chief but contemporarily it is more likely to be a traditional medical practitioner or elder. Within three to six months of leaving the cave the barren women is expected to be with child. One interviewee claimed she was born as a result of her mother staying three nights at Kyumbe (MC/Nr24/July 2004).

Sieny is also associated with a local area of forest that is indigenously referred to as 'Sieny Forest' or 'Valley of the Trees' but which official cartographic and political enterprises refer to as Foo Forest Reserve. The forest belt on Mt Kilimanjaro is a protected area. Most of the forest sits within the Kilimanjaro National Park boundary and therefore is theoretically inaccessible to local people. Sections of forest lie outside of the park boundary and rights of access and tree-felling in these areas are prescribed to various groups (MC/Fo37/July 2004). The area of forest within the valley that separates the villages of Foo and Nronga is believed to have spiritual potency. The belief in the sacredness of the forest is not uncommon within the wider region indeed others have documented the sacred forests of the Pare (see Sheridan 2002). Many local informants report stories of the big trees to be found there that are as old as *Ruwa*. Again the cosmological theme of historical existence and previous temporal orders being equated with divinity is apparent. This form of cultural perception that homogenises the distant past with the sacred, including the performance of related rituals, are associated with relics. These rituals and perceptions are relics of the past that provide inspiration and motivation for future/present behaviour and practice based on past achievements and potency. Another feature of the Sieny area is the myth of the iron hut. Numerous interviews document this phenomenon (e.g. MC/Ud12/July 2004; MC/Nr6/July 2004). The feature is believed to be the generator and origin of the tribal-ethnic strength and

prosperity. In this capacity it protects and bestows on the Wachagga resolution, insight and wisdom (MC/Nr9/July 2004). Multiple informants could give instruction as to the approximate situation of the hut but after numerous attempts to locate the phenomenon the approach was aborted. Not least because informants believed the hut was situated in different places. Others later related the transient and ethereal nature of the hut and therefore reasoned that it was simply not there during my investigations (MC/Nr33/July 2004). Descriptions of the feature varied considerably but most informants concurred that it is exceptionally large and is constituted of 'iron-rock' (MC/Nr6/July 2004). A similar feature of the memoryscape was encountered in Rombo (see below).

Figure 19.1. Photograph of Daraja la Mungu from position of high relief

Figure 19.2. Photograph of Daraja la Mungu from on the feature

Figure 20.1. Photograph of Nkyeku cave

Figure 20.2. Photograph of Nkyeku cave illustrating the markings of significance

Three other key features of the local memoryscape were highlighted through interview in this locale. It is interesting to note that they all fall outside the accepted geographical and administrative limits of Sieny. Nevertheless they were repeatedly discussed in relation to Sieny and this may relate to a sense that water attributes, which are sacred and integrative, locally cohere. Mura wa Ngumbe (Kimefu Waterfall) derives from the Namwi River and as such the relation to Sieny is clear. Precolonial irrigation furrows are still in operation and distribute water from the site throughout the local banana groves. The water and local vegetation from Mura wa Ngumbe is believed to possess medicinal properties. Diseased cattle are brought to the site to drink and traditional medical practitioners frequently harvest materials from the area to use in their medicine (MC/Nr6/July 2004; MC/Fo2/July 2004). Another feature is Kukeleweta Chemka (**figure 21**). This is an area of naturally-heated springs near the village of Kukeleweta which is 10km from the Hai district administrative centre of Boma ya Ng'ombe. The area is 28km from Sieny. It is thought that the feature is a recent inclusion within the local sacred memoryscape that has become more compelling with the increasing levels of migration and mobility of local persons. Chemka means 'hot spring' in

Kiswahili. Many interviewees thought the place was the site of a supernatural force or spirit (e.g. MC/Fo6/July 2004). This force was thought to trick people by enticing them into the water to their death (MC/Nr17/July 2004). It is probable that this belief derives from stories of the death of a young woman a decade previously. Depending on the source at Kukelweta the death was thought either to have been drowning linked to inebriation (BN/Ku3/August 2004) or else a crocodile kill (BN/Ku5/August 2004). Although it should be highlighted that it is rather unlikely that crocodiles were the culprits as their presence in the warm water would be somewhat abnormal (see Huchzermeyer 2003). In Machame various interviewees recalled efforts made by themselves and earlier generations to travel to the springs to collect a mineral called *magadi* to feed diseased livestock (MC/Fo35/July 2004; MC/Nr6/July 2004). This underscores the relationship between the supernatural and environmental subsistence. The sacred is an inherent component of all habitual existence. It was surprising to discover that at the village of Kukeleweta the inhabitants thought the area quite interesting but devoid of a sacred dimension. For them the springs were simply a source of tourist income (BN/Ku2/August 2004).

Figure 21. Photograph of Kukeleweta Chemka

Figure 22. Photograph of Lake Nyumba ya Mungu

Lake Nyumba ya Mungu (-3°80' 37°47') is another feature symbolically and supernaturally associated with Sieny whilst being geographically distant. The vast water source sits between between the Pare Mountains and Kilimanjaro about 110km from Foo village. The villagers on the Machame transect articulated various beliefs about the place. The locals knew that it was once widely believed that Nyumba ya Mungu was the dwelling place of *Ruwa* (MC/Ns11/July 2004). The place name actually translates as 'House of God'. It is believed that the lake is powerful and was the dwelling place of spirits that are associated with the divine. The interviewees thought the place sacred and some reasoned this spirituality by recounting the story of this enormous stone in the centre of the lake where *Ruwa* once sat. The evidence was the stone and the informants failed to see how such evidence could be interpreted in any other way (MC/Fo19/July 2004). One interviewee reported that if you are accepted into Nyumba ya Mungu (**figure 22**) you will be able to "meet and listen to the past people" (MC/Nr8/July 2004). The site is conceptualised by some as a permeable threshold between the distant ancestors and the living. It is interesting to note that the lake is a major source of Hydro-Electric Power regulated by TANESCO. Due to the perceived political instability, insurgency and expansive agendas of neighbouring countries and the potential attractiveness of such a plant as a strategic target, there are extensive security measures surround the site. Access is limited and regulated (although villages established on the periphery of the lake have extended rights), photography of any form is banned, permits are required for entry, and the area is fenced and patrolled near every one of the all-weather access roads. These measures have obviously exacerbated beliefs in the sacredness of the area (MC/Nr22/July 2004; MC/Fo27/July 2004).

KISUMBE RESERVATION

Upon initial observation Kisumbe Reservation (-3°16' 37°31'), also known as Yenunyi, seems an unremarkable area in Mshiri village. It is a limited wooded coppice devoid of agriculture or grazing (see **figure 23**). The area is reserved because of the numerous irrigation furrows and natural water channels that diverge and confluence there. It is vital to keep the area unpolluted to prevent harmful repercussions down-channel (MR/Ms14/August 2004). The reservation is also important locally for other less prosaic reasons. Indeed the area is imbued with much mythic and historic agency and as such is an exemplar of the complexities of local memoryscapes. According to local tradition the area was originally arid wasteland but filled with water during a period of intense drought after locals performed many rituals for *Ruwa*. The reservation, an endowment from *Ruwa*, appeared from nothingness (MR/Ms1/August 2004). The term 'Kisumbo', from which the place name Kisumbe is derived, means 'a gift from someone'. The area is also associated with the existence and actions of the Kisumbe Snail. The cultural

narrative of the snail is a vital component within the local memoryscape that facilitates the interpretation of historic and contemporary power relations between the rival patriclans of the area. In the past Kisumbe was the dwelling place of the snail. Marangu, as indeed land throughout most of *Uchagga*, was the site of inter-generational patriclan warfare. Two of the major patriclan protagonists in the vicinity, historically and contemporaneously, are the Mtui and the Moshi/Marealle patriclan. The oral tradition notes that these two major patriclans were always engaged in a bloody rivalry with a constantly shifting local balance of power (MR/As9/July 2004; MR/Ms4/August 2004). Since the Moshi/Marealle, throughout the German and British administrations, occupied both the positions of Mangi Mkuu (Paramount Chief of all the Wachagga) and the Marangu Mangiate (Divisional Chief) the struggle clearly ended with the Mtui in the unfavourable position. According to local mythic narrative the Kisumbe Snail was very much involved in this precipitated eventuality.

Local tradition notes that during the later patriclan wars the Moshi/Marealle outnumbered and strategically outmanoeuvred the Mtui. This imbalance was redressed when the Mtui began taking their dead and wounded to Kisumbe for treatment - for the snail would give life back to the dead warriors and heal those that were wounded (MR/Ms4/August 2004; MR/Ly16/August 2004). It is, of course, possible that the healing properties of the Kisumbe locale may relate to the abundant clean water that could have been put to cleansing and medicinal uses. The area is still thought to hold healing properties especially for males (e.g. MR/Ms19/August 2004). The myth of Kisumbe meant that many informants concurred that the dead would regain life once the supernatural snail had crawled over or 'licked' their lifeless bodies. Obviously with this newfound asset the tide in the war changed substantially.

> "The Kisumbe Snail was the property of the Mtui for it would bring life back to the warriors so they would fight again. The Snail made the Mtui strong and with it the Mtui repeatedly defeated the Marealle" (MR/As11/July 2004)

This situation was not historically ordained though as the power relations were later re-orchestrated. Unfortunately for the Mtui their patriclan was infiltrated by a beautiful Moshi/Marealle slave girl who would betray them. Eventually after many years of dutiful service the girl was trusted with the location of Kisumbe and the task of feeding the snail, which ate only honey and milk, and she supplied this information to the Moshi/Marealle. They immediately dispatched warriors and killed the snail (MR/As26/July 2004). This catalysed a final shift in power and the Mtui lost their dominance. It is worthy of note that the place name 'Ashira', one of the local villages, actually translates as 'dead snail' for it is where the corpse of the Kisumbe Snail was moved to after its death (MR/As2/July 2004). The narrative of the snail is given much local currency and to the present day the Mtui and Marealle clans never intermarry. This is linked to the historic rivalry but also to the understanding that such unions would eventually lead to betrayal and disloyalty (MR/Ly8/August 2004) and bring suffering, premature death, still-birth and other negative consequences (MR/As24/July 2004). The fact that the Mtui patriclan never employ servants also relates to these events (MR/Ms36/August 2004). The Kisumbe Snail is more than a local myth it is a belief based on memory – the collective memory of the patriclans which is inscribed upon the landscape. The landscape acts as a priming device that repeatedly expresses the past.

Figure 23. Photograph of the Kisumbe Reservation

Associated with the historical sacredness of the Kisumbe Reservation is the nearby Kinukamori Waterfall (-3°17' 37°31'). For view from the top of the falls see **figures 24.1 and 24.2**. The relation derives primarily from the cultural perception of water that possesses an inherent force of connectivity but also from the fact that the two sites correspond in their relation with the Mtui patriclan (more below). The visually striking falls are the site of historic agency. This action forms part of the local mythic memoryscape. All indigenes are familiar with the legend of the ill-fated Makinuka. Makinuka was a young woman who, tradition holds, lived during the reign of Chief Rina of Kibosho. This woman was involved in unchaste relations with a local man. Penalties for such behaviour were particularly harsh and the two of them were fortunate not to be caught. Interview data gathered along the Marangu transect informed that the typical penalty for inappropriate sexual conduct at this time in both Marangu and Kibosho was death. This ordinarily entailed tying the two wrongdoers together, facing each other on the ground, and then driving either spears and/or stakes through them both and into the ground. The two would then be left to die (MR/Lb/15/August 2004; see also Dundas 1968: 296-7). Nevertheless Makinuka became pregnant and her swelling began to testify to her recent activities. Her mother cast her from her home. Makinuka decided to kill herself by throwing herself from the top of the falls. Upon reaching the top she reconsidered her action and decided to return home and plead for mercy. Just as she was about to move she became aware she was being stalked by a leopard and upon nervously backing away fell over the falls to her death (MR/As21/July 2004; MR/Ly8/August 2004).

Figure 24.1. Photograph looking-out from the top of Kinukamori Waterfall

Figure 24.2. Photograph looking-back from the top of Kinukamori Waterfall

The falls have been recently exhibited in such a way as to maximise tourist potential. Life-size concrete casts of a leopard and a woman have been erected and the adjoining *Chagga Cultural Museum* is seeing a steady flow of tourist visitation. With or without these recent amendments the waterfall serves as a moral primer. When one who is familiar with the local traditions views the waterfall one recalls either consciously or unconsciously the story of Makinuka. Moreover the falls retain a potency related to the systematic agency of the past. Traditionally the uncircumcised dead were thrown off the top of waterfalls to be consumed by the animals (MR/Ms9/August 2004). It is possible that this myth may originate in such practice. The name of the falls is derived from its past ownership. It was the property of a Mtui patriclan elder called Kimori. Indeed the Mtui clan are often referred to as the Kimori or Mmori clan. Hence Kinukamori relates to Kimori Mmori. The site also has other less pleasant connections with the Mtui clan. It was the site where punishments were delivered to convicted clansmen during the period of the precolonial chiefdoms. Moreover it is where Nathaniel Mtui was murdered.

Nathaniel Mtui was a powerful local who worked for the German colonial administration. He was one of the candidates who many thought might be appointed Paramount Chief of all the Wachagga – a position that was eventually occupied by the Marealle clan.

The falls are revered as possessing sacredness and as the dwelling location of *Ruwa*-spirit. This spirit is believed to be a latent power that should be respected. This notion of dormancy is also applied to the mountain (MR/As4/July 2004). The falls are also important as an area to collect leaves and shoots of *mriri* trees. These trees are believed to posses medicinal qualities that when properly released will cure various ailments including mumps. It is reported that if one suffering from mumps circles the tree three times, singing *mriri shimba ndoyo* translated as 'mumps get out of me', whilst wearing neck gourds s/he will be cured of the ailment (MR/As16/July 2004). This religious interpretation of the falls has recently been exacerbated with observations of Hindu's throwing the cremated ashes of their dead from the top of the falls (MR/Ms28/August 2004).

KIFINUKA

There are two kinds of spirits that occupy places, features of the landscape, and more rarely people and animals. Well-meaning spirits and energies are known as *Ruwa* and are associated with God. *Ruwa*-spirits are fractals of God but at the same time can be dislocated and independent of the higher supernatural whole. Bad spirits are referred to as *mbarimu* in Kihorombo (*mzimu* in Kiswahili). There are also *Majini* spirits which are relatively recent additions to the local memoryscape. The area and features in Maharo known as Kifinuka (-3°12' 37°32'; see **figure 25**) embodies a myriad of all these types of spirits. The area sits on the boundary of two villages – Maharo and Shimbi. The area formerly contained a large body of water that was drained by the elders after the *mbarimu* within it became too strong. Traditionally the water pool was where circumcision parties were taken so that one of the boys could be sacrificially drowned in order to bond the party and confirm the groups' initiation into adulthood (MH/Mh31/August 2004). The lake was drained though after it started "taking many boys of its own accord" (MH/Mh4/August 2004). The intervention of the elders reduced the negative effects of the *mbarimu* in the area. The area is also believed to possess remnant powers of a positive nature because of its involvement in much historical ritual action and intercommunion with the ancestral plane. The locality was involved in processes of making people through initiation, sociality and ritual action, and so due to this living memory of the recent ancestral past a shared communal identity is embodied and transcribed on the place. This useful potency can be accessed through sacrifice and offering (MH/Mh17/August 2004).

On the periphery of Kifinuka lies the forest which is involved with local sacredness. When one intends to engage the assistance of the ancestors and/or *Ruwa* one must perform rituals in the sacred forest (MH/Mh14/August 2004; MH/Mk39/September 2004). The highly ritualised slaughtering of the forest must be performed in duplicate and this binary association is stressed. Firstly two animals must be sacrificed. One must be slaughtered and consumed immediately whilst the other must have its eyes gouged out and then released so that it might wander. Moreover the slaughter is usually repeated some time after the initial episode and the ritual transaction is not considered complete until the secondary slaughter has been conducted (MH/Mk28/August 2004). It is possible that the forest location hides the ritualistic activities from the local church leaders although most elders who were interviewed confirmed that these things had always been done in the forest. Of course whilst agricultural and subsistence demands on the land were less the forest was more widespread. There is an adaptive dimension to the types of offerings that "give to the forest" (MH/Mh14/August 2004). The offerings which traditionally related to requests for safety and prosperity may well have reduced the number of attacks by wild

animals by providing easy prey. This ecological adaptivity could well account for the longevity of such practice and with time the rituals gradually became more elaborate.

Another key feature of the indigenous mythic memoryscape prevalent around Kifinuka is that of the spirits of Kibo and Mawenzi. According to mythic tradition there used to be two spirits (others called them angels, gods or forms of *Ruwa*) that respectively dwelled on the Kibo and Mawenzi peaks. They will be referred to herein as spirits as that was the most common descriptive term encountered (e.g. MH/Mh12/August 2004). In the past these spirits used to prevent people reaching/climbing the peaks or higher locations of the mountain with their supernatural powers and tricks. Most informants concurred that the spirits are no longer apparent – hence people are climbing the mountain. Nonetheless the twin peaks still have remnant powers from their period of occupation. The landscape embodies the potency of the spirits and their past agency.

> "Both Kibo and Mawenzi are still powerful from that time. They are alive with the forces of the spirits that lived upon them. If you go up the mountain to the peaks you will encounter things that you might think cannot happen and these things are the results of those angels (MH/Mh12/August 2004)

These assertions were extrapolated with revelations of places high up near Mawenzi peak where both hot and cold water are found. This comes from "within the place" (MH/Mh19/August 2004). Generally it is also noted that most of the rivers and water channels that run down-mountain through the *kihamba* emanate from an area between the two peaks. This emanation is believed to be dependant on the latent powers of these spirits. Moreover other commentators noted that during the dry season there are things that glitter and shine from Mawenzi which are explained in terms of these supernatural entities (see MH/Mh34/August 2004).

These twin spirits are also deemed responsible for other supernatural agency and properties on the mountain. The climatic and environmental conditions above the forest belt and higher are understood as being a consequence of these former supernatural entities. In fact the sheering cold and the snow are so anomalous to anything else encountered in the bush or lower slopes that they cannot be explained away in any other terms. Moreover there are certain special places high up on the mountain. When you reach these places you must not talk otherwise thick clouds smother you and unrelenting rain will fall. These conditions are considered extremely dangerous because becoming lost would be unavoidable and death would be highly probable (MH/Mh19/August 2004). Dundas (1968: 37) records a similar theme regarding the onset of blindness high up the mountain. Again supernatural agents were thought to dwell within a cave full of ivory –

Figure 25. Photograph from position in Kifinuka

an elephant graveyard – which would take away one's sight. This could be an earlier articulation of the becoming lost narratives. It has been logically suggested that these myths may relate to cases of snow blindness (Dundas 1968: 37). The numerous cave complexes above the forest belt are also considered to be particularly dangerous due to supernatural entities embodied therein. The *mbarimu* that occupy them tempt people inside and they are unable to leave. There were numerous stories of *muzungu* entering the caves and becoming lost. These same *muzungu* were the source of much amusement for it was remembered that they did not heed local advice after seeking it (MH/Mk9/August 2004). The caves are very much conceived as being alive in an anthropomorphic sense.

> "Those caves high up the mountain are very hazardous. They entice many people inside them to their death. The caves have desire and they have an appetite for consumption" (MH/Mh22/August 2004)

It should be noted that the existence of these places, powers and properties are generally accepted although

some interviewees suggested that rather than all being attributable to the twin spirits of Kibo and Mawenzi other more localised spirits were responsible (e.g. MH/Mh5/August 2004).

A further supernatural force in the area is *Mba ya Mungu* or 'House of God' in the local dialect. The myth is similar to the iron hut stories of Machame (above) and they probably share a common origin. *Mba ya Mungu* is an enormous free-standing rock high up in the forest. The rock befriends hunters because it protects them. The rock is hollow and is the size of a local dwelling. It can comfortably fit a moderately sized hunting party (three-four people) in it. During the day the rock has a large opening that permits the hunters in but at night the opening closes to protect everybody inside from wild animals (MH/Mk19/August 2004).

> "High up there is *Mba ya Mungu* the protector of the hunters. The door is controlled by God. At night it will close so that the hunters can hide in it and be safe" (MH/Mk40/September 2004)

Despite efforts during the fieldwork to locate *Mba ya Mungu* it remained unfound. According to numerous informants it only appears when it is needed by local hunters.

"You cannot find *Mba ya Mungu* – it will come to you if you are in need of protection this is why when you see it you must get in for there is some danger coming for you" (MH/Mh34/August 2004)

CHAPTER TEN

CONCLUDING REMARKS: MEMORYSCAPES AND ARCHAEOLOGY

The memoryscapes described herein through collation of interview material share an area of commonality or mutuality. The mythic teleological construal demonstrates gross particularism. In their respective contexts the various landscapes exhibit different metaphoric and physical features. In this sense the creation and articulation of mythic memoryscapes are expressions of cultural attunements and being-in-the-world. The perception of the landscape is culturally-informed and the religious or supernatural elements of cultural existence are at once both encoded onto and read from the environment. The sacred landscapes of *Uchagga* inform and choreograph the being-in-the-world of those who dwell within it.

There are numerous avenues for further investigation and research concerning the contemporary and historic memoryscapes of the Wachagga. In many ways this research venture has enjoyed – and at the same time been inhibited by – the freedom of exploring indigenous cultural landscape perception without being able to engage with very much previous research on similar themes. Further research into the recapitulation of traditional religion in the contemporary worldview would be rewarding. It would also be hugely advantageous to increase the comprehensiveness of the ethnographic survey. In particular the description of memoryscapes of indigenous groups from the more remote and inaccessible areas of the mountain would enhance those already collected. Moreover ethnoarchaeological projects conducted at sacred sites might prove rewarding. It would also be sagacious to note that the subsistence and dwelling of the historic and contemporary indigenes of *Uchagga* may represent an apposite archaeological case study for those researching into the origins of agriculturalism and pastoralism. The indigenous farmers have expertly exploited the productive landscape in conditions perhaps similar to those facing earlier groups who abandoned gatherer-hunter subsistence. Most ethnographic analogues erroneously highlighted in such discourses are usually geographically remote marginal groups struggling to subsist. Clearly the later gatherer-hunters and early agriculturalists would have sequestered bountiful and resource-laden landscapes.

The quantity of crucial information that is unknowable about the past is significant if one considers interpretations about past landscapes. In terms of the processual cognitive, optimisation and distribution models such studies highlight the lack of human place in models of theoretical space. Moreover when one considers the applications of phenomenology to landscape archaeology these considerations highlight the elements of human experience that are beyond interpretation. These considerations have been considered in detail elsewhere and so will not be repeated here (see Brück 2005). This is not to suggest, however, that phenomenological methods in landscape archaeology have not been a positive addition to the battery of approaches that can be deployed. Indeed these pioneering approaches to landscapes have taken us well beyond the sterile boundaries of *terra incognita*. It is worth recalling how these methods conceptualise place.

Thomas (1996, 1999, 2004), for example, notes that there is an implicit reliance on the ontological primacy of the material world underlying the majority of the processual and postprocessual archaeological thinking and uses phenomenology to critique the Cartesian underpinnings of such reliance. Thomas argues that the world can never exist in isolation from being-in-the-world. He discusses prehistoric places, monuments and artefacts and considers how these can become bound-up or constitutive of the person and in doing so deconstructs the categorical distinction between self and other (Thomas 1996: 78-81). The world is found through its historicality and its disclosure is thus dependent upon prior individual and cultural experience (Küchler 1993). It has been pointed out that this is why local and regional histories feature more prominently in the work of Thomas than in those of other phenomenological archaeologists (Brück 2005: 50).

In contrast Tilley (1994, 1996, 1999, 2004a, 2004b) reengages with the qualitative aspects of landscape by exploring the ways in which social and cultural meanings are bound up with places. This is accomplished through methods that recognise that human experience and interpretation are bodily mediated. Space is situated and contextual (Tilley 2004: 11). The physical engagement of the human body with the material world is central to experience i.e. bodily movement through space makes such space meaningful because particular ways of viewing the world are availed. Tilley argues that to understand the material remains of the past it is necessary for the archaeologist to document their own physical engagement with these spaces as they move through them (Tilley 1994: 73; 2004a: 27-8). As Brück (2005: 48) notes this method involves describing (through words and photographs) the views from particular locations, the order in which different spaces and locations are encountered, the way monuments or their compositional elements reflect the physical landscape, and one's own bodily experiences of features. Thus the encounter between the archaeologist and the archaeological material in the present provides understanding as to the ways past people encountered this material because the human body is the shared medium of experience. Tilley (2004a: 225), methodologically aligning with Gadamer, stresses his prejudices and biases in his "creative responses" to the material world. In such a vein other archaeological literature has included personalised diaries and accounts of experiences (e.g. Bender *et al* 1997; Hodder 2000). Approaches similar to Tilley's have been developed by other archaeologists to inform the available evidence (see Brophy 1999; Cummings 2002a, 2002b; Fraser 1998; Richards 1996; Watson 2001). While other archaeologists have applied phenomenological approaches to everyday practice and inhabitation, rather than monumentality and landscapes, noting that the everyday routines constitute social relationships (Barrett 1994; Edmonds 1999; Gosden 1994; Pollard 2000). These archaeologies concern landscapes of routine which sediment themselves into our being rendering them part of ourselves (Brück 2005: 62).

Many researchers have replicated the methodologies of phenomenological archaeology to produce their own descriptions of experiencing landscapes and monuments, yet it is unclear to what extent these studies have added to the understanding of the past (Brück 2005: 54; Insoll 2004a: 86). The primary criticism of phenomenological archaeologies concerns whether the embodied response of an archaeologist in the present can ever approximate the actual experience of people in the past (Brück 1998; Hodder 1999). And if not are such exercises forms of academic imperialism colonising past worlds with the experiences of the present, in particular those primarily of the white, heterosexual, Anglo-Saxon male (Gosden 1996; Weiner 1996). How can there be some commonality of experience? In answering this question it should be noted that on the one hand there is considerable variability in the physical attributes of the human body such that different people (or beings) might be expected to experience the material world in significantly different ways e.g. in terms of sex, age, health, dis/abilities and so forth. On the other hand the body is meaningfully produced in the social realm and therefore embodied experience is shaped by cultural attunements. Thus the experiences of the 'same' material conditions can, and would be expected to, vary both within and between social groups (Brück 2005: 55). Moreover movement and actions within or through landscapes, places or monuments will take the form of culturally learnt bodily practices that are socially meaningful and impact upon

their experience e.g. walking, climbing, kneeling, even viewing (Bourdieu 1977: 94).

There are further potential pitfalls for phenomenological archaeologies and the pasts they recreate. Phenomenological accounts often demonstrate ethnocentric naivety in their failure to recognise that cultural attunements condition experience at the level of the individual (Laughlin 1999: 459). One assumes that most archaeological theorists have not in the main been trained to cultivate phenomenologically productive mind states (Clack 2004: 58). Moreover, archaeologists applying phenomenology to the material past run the risk of being 'phenomenologically unsophisticated'. This 'unsophistication' does not necessarily only equate with insufficient depths of knowledge (Gosden 1996: 23-4) but also an absence of more mainstream philosophical contestation. Numerous archaeologists have noted that the visual field, to the exclusion of the other significant senses, has been overemphasised in phenomenological applications (Cummings and Whittle 2003; Mills 2000). Despite these criticisms, however, it is clear that the condition of humanity is composed through relations, memories, histories, meanings and symbols. As Tennyson implied, the person, through their experience, is part of the landscape and the landscape is part of the person. The relationships between memory, experience and landscape are all involved in making and unmaking the person.

The motivating rationale of this ethnographic research was the theorisation of some issues involved in landscape archaeology. The study augmented such appreciations by theorising the complex environmental relations of a contemporary culture. Such theorisation enables archaeology to fully recognise the wealth of human experience that is archaeologically unknowable through a worked ethnographic project. As has been mentioned this is not to contend that archaeologists have been completely ignorant of such insights but rather that without a deliberate comparative exercise the scale of the omission has conceivably been underplayed. It is important to note, however, that landscape archaeologies can never fully 'recreate' the past. Any past will always involve a dialogue between subjects in the present and those in the past. This dialogue is mediated through the archaeological record. The archaeologist interprets the past and is concerned with accessing the knowable. It is vital to remain conscious of the unknowable and the ways in which such might significantly impinge upon the knowable and the conditions of possibility.

FIELDWORK METHODOLOGY

Fieldwork Details

The fieldwork was conducted between June and September 2004 in the Tanzanian district of Kilimanjaro. The fieldwork primarily consisted of the collection of oral historic traditions and ethnographic data through interviewing and participant observation. In this sense it is an example of ethnographic research defined as the interactive participation of researcher and subjects, overtly or covertly, for a period of time collecting whatever data becomes available to illuminate the research (Hammersley and Atkinson 1995: 1). These techniques have been used in the region with some success in the past (e.g. Stahl 1964; Howard and Milliard 1997; Moore 1986; Tagseth 2000).

Aims and Objectives

The primary aim of the study was an investigation into the memoryscape (and in particular the environmental relations) of the Wachagga. This study was to be conducted through an assessment of the reported clan traditions and origin myths that posited the Wachagga came from the mountain (Stahl 1965: 37; Odner 1971: 131), or were dropped from the sky through divine action (Lema 2002: 44). The interview stage of the project had the objective of verifying these origin myths and their permeation and proliferation within the local culture. The interviews were also planned with the aim of verifying how long the oral history dictates the Wachagga have lived on the mountain which is generally accepted to be 450 years (Moore 1977: 5). From a literature review it was apparent that what historical work exists contradicts the previously collected oral narratives for the locale. It has been proposed that the Wachagga slowly migrated and settled in an up-mountain direction due the unfavourable agricultural conditions found in the bush and at higher altitudes (Wimmelbücker 2003: 46-7). Nonetheless the oral historic traditions are corroborated with some archaeological survey data that suggests the middle and upper zones contain more sites than the lower zone (Odner 1971: 134).

Mapping

Mapping and survey of the area is severely lacking. The recent national census in Tanzania, which professes to have spent considerable time and resources on the mapping stage of their project, has not published any such maps with the census results or provided copies to regional authorities. This project therefore had to use the 1: 50 000 and 1: 100 000 OS map sheets that were based on aerial survey material collated in 1965 by the British Government Ministry of Overseas Development. This was subsequently updated with information forwarded to the Tanzanian Surveys and Mapping Division which was published in 1989. No field checks were made of this data. Many of the features on the map sheets are therefore misleading or inaccurate. This mapping situation had severe implications for the study. Considerable time was spent calibrating and triangulating GPS readings with compass work and relating these results with the available map resources. All maps provided herein are hybrids of information taken from the aforementioned map sheets and personal survey work with handheld equipment.

Transects

During the study three transects were positioned, one in each of the districts that neighbours/borders Mount Kilimanjaro – Moshi Rural, Rombo and Hai. Transects were located with access and prudent resource-utilisation very much in mind. These linear transects ran in an up-mountain direction from the lower slopes towards Kibo peak (see **figure 26**). Access was restricted to the periphery of the Mount Kilimanjaro National Park due to permit and research controls. This had little implication for the ethnographic and oral history study due to the zero inhabitancy rates in the park. It should be remembered that despite being useful for analytical and descriptive purposes the village entity as a unit of households is not locally relevant (Moore 1986).

- The Marangu, Moshi Rural (MR) transect encountered four main village settlements – Ashira, Mshiri, Lyasongoro, and Mbahe. Ashira, Mshiri and Lyasongoro villages lie in the Marangu East Ward, and Mbahe village lies in the Marangu West Ward, Marangu Division, Moshi Rural District, of the Kilimanjaro Region. These wards are positioned south of the Kibo/Mawenzi peaks.

- The Maharo, Rombo (MH) transect encountered the village of Maharo, in the Maharo Ward, and Makiidi village in the Makiidi Ward. Both these wards fall under the coordination of Mkuu Division, Rombo District, of the Kilimanjaro Region. The wards are positioned southeast of the Kibo/Mawenzi peaks.
- The Machame, Hai (MC) transect encountered farmsteads from four village settlements – Foo, Uduru, Nshara and Nronga. Foo, Uduru and Nshara are villages administered by the Machame Kaskazini Ward. Nronga village is in the Machame Uroki Ward. Both the Machame Kaskazini and the Machame Uroki wards are in the Machame Division, Hai District of the Kilimanjaro Region. Both wards are positioned southwest of the Kibo/Mawenzi peaks.

Figure 26. Map showing locations of the three interview transects on the slopes of Mount Kilimanjaro

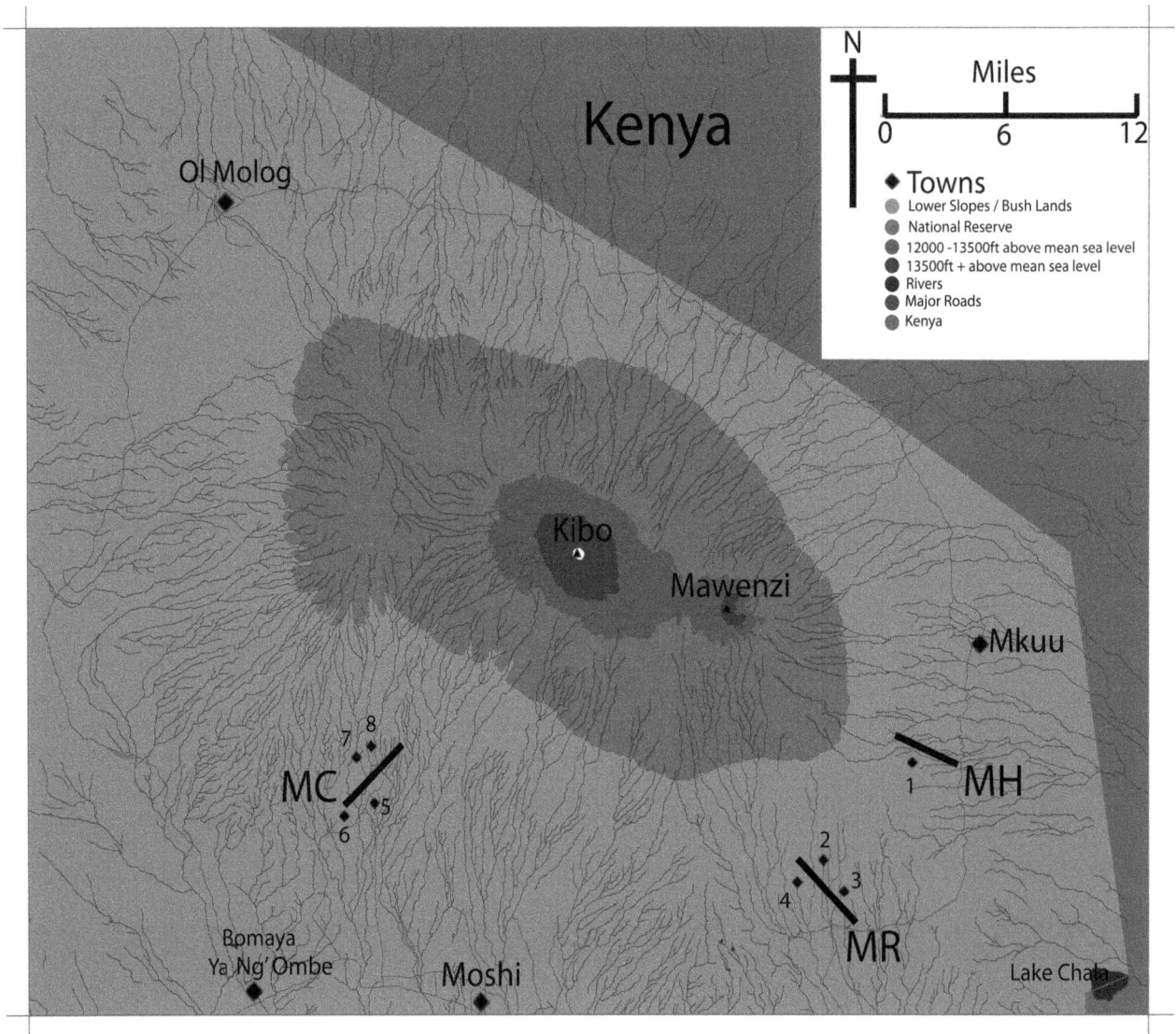

Anonymity Issues

Religion and ritual behaviour are politically, socially and economically sensitive issues in sub-Saharan Africa and the Kilimanjaro region is no exception. There is considerable recent anthropological literature on these dimensions of religion in similar environments (see Abbink 1998; Campbell 1999; Gatwa 1999; Greene 1997; Jambai and McCormack 1996; Kalu 1999; MacGaffey 1990; Rich-Dorman 2002; Tokunbo-Williams 1997; White 1993). Missionary activity in the area during the colonial administrations was immense. As Vansina (1990: 239) highlights missionary agency was utilised as a justifiable element of the 'civilising' imperialist project. The colonial actors frequently invoked the principles of civilisation and modernisation to obliterate facets of the traditional. It is noted, however, that the colonially imposed structure of customary law was utterly alien to the character of the former tradition. Indeed "[c]ustomary law was the headstone on its grave" (Vansina 1990: 239). Perhaps this overly devalues the concessionary nature of customary structures and the integrative agenda of the colonisers. Fortunately the project of obliteration was not completely successful and uncountable manifestations of traditional cultures were merely re-categorised as deviant due to their covert performance.

The various churches represented in the region maintain considerable spiritual, economic and collective influence over the indigenous inhabitants. In appreciation of this situation care was taken to minimise any negative impact the research could create. Copies of permits, letters of introduction, and research clearance were made available for interviewees to scrutinize and takeaway. Examples of these documents can be seen below. All interviews were initiated with declarations concerning the research and its practitioners. Some interviewees needed considerable reassurances as to our non-employment with any government ministry, religious and/or military organisation, and others still refused to answer specific questions. A very small minority refused to be involved with the project entirely. Furthermore due to the sensitivity of some of the issues discussed many informants declared their wish to maintain anonymity in any publications derived from the research. Thus in line with other projects executed in the area (e.g. Hasu 1999) the identities of informants have been safeguarded to the best of the author's judgement.

Interpretation and Translation

All interviews were conducted in the local Kichagga dialect and translated into English. Occasionally the interviewee would respond to questions in Kiswahili or English when this occurred responses were recorded. Subsequently the question was rearticulated and the answer requested in local dialect. This practice was adopted to minimise the interpretive loss of meaning inherent within acts of translation. With such a translatory method only one interpretive process was undergone. The same translators were used throughout the interviews of each transect with the linguistic reasoning that such a protocol would offer control and uniformity to the process. Due to dialectic considerations differing translators were required at each transect as local Kichagga dialects are mutually unintelligible. The translators were therefore local and thereby 'insiders' (see Devereux and Hoddinott 1992). Although their ability to communicate with all parties made them 'outsiders' in the sense of socio-economic grouping and resource access. This seemed to promote a good rapport with the interviewees who were comfortable with the presence of the interview team but were not too familiar as to create methodological problems (see Tagseth 2000 for more detailed discussions of these considerations). Interview data was written-up as soon as possible after leaving the field each day. Transcriptions were either based on recorded audio-tape or field-notations made by the author during the interview.

- *Translation and Interpretation on the Marangu (Moshi Rural) Transect*

Translation in Ashira and Mshiri was conducted by FM (sub Vice-Chairman of Mshiri village) with the occasional presence of RM (Mshiri Village Chairman). At Lyasongoro and Mbahe villages FM continued the translation process accompanied by RSM.

- *Translation and Interpretation on the Maharo (Rombo) Transect*

Translation in Maharo and Makiidi villages was conducted by PS (retired accountant) with the assistance of PNS (school teacher).

- *Translation and Interpretation on the Machame (Hai) Transect*

Translation in Foo, Uduru, Nshara and Nronga was conducted by RRS (farmer), CU (school teacher) and JSM (retired school teacher). Occasionally interviews would be conducted in the presence of JTM (Machame Kaskazini Ward Administrative Secretary).

Methodology

Households on farmsteads bisected by transects were targeted for interview. Where it was practicable interviews were conducted outside the household in question. This was preferred to the alternatives because it was assumed the interviewees would feel more comfortable in a familiar domestic setting and communications with other researchers in the area had informed of local hospitality. This interview practice is generally accepted as the best strategy to gain trust (Hammersley and Atkinson 1995: 150). It was also thought that areas of the local landscape that may be of potential interest to the study would be more accessible. Due to the low mobility of the indigenous populations it was always possible to interview houses that were inhabited. Interviews followed a structured questionnaire format with follow-up questions being asked when required. The elders of the household were prioritised for questioning. This prioritisation was necessitated for various reasons. Firstly because of the respect the other family and village members afforded the elders subsequent interviewing of other parties was much easier. Secondly because of their age it was assumed that the elders had access to, and were familiar with, more elaborate memories and myths. Thirdly the elders were traditionally the ones who remained closer to the farmstead throughout the day due to mobility, security and labour-requirement issues. The interviews established oral traditions for the areas.

The oral narrative and tradition should be considered invaluable anthropological and historical instruments for the investigation of the past and the present. Oral traditions are historical phenomena that exist in all cultures. They are "fundamental continuities which shape the futures of those who hold them" (Vansina 1990: 258). Generation and maintenance of traditions involves the mediation of perception, emotion and memory. Moreover experiential interpretation, commentaries and iconatrophy, improvisation and performance are all inherently related to the reproduction of the tradition and provide insight into cultural perspectives, historicity and comportment (see Vansina 1985).

> "[T]raditions are self-regulating processes. They consist of a changing, inherited, collective body of cognitive and physical representations shared by their members...[that]...give meaning to changing circumstances in the physical realm" (Vansina 1990: 258)

In establishing traditions numerous interviews were conducted (n=316) at various times during the fieldwork along the three transects. At Machame (n=111), Marangu (n=102) and Maharo (n=103) people were generally very keen to be involved and genuinely interested. This obviously consumed enormous quantities of field time. The data recorded and represented in the study displays a distinct age bias because of the AIDS-related deaths of the sections of the population who were of reproductive age, the high proportion of males who work in the local urban centres for considerable periods at a time, and high infant mortality. In the study elder or *mzee* status was afforded to an individual over the age of sixty. This was cross-referenced with the local translator or guide who would confirm if an interviewee was locally considered an elder or not. The proportion of elders interviewed on each transect varied. At Machame (n=62, 55.8%), Marangu (n=71, 69.6%) and Maharo (n=59, 57.3%) elders were interviewed. The interviewee gender balance was slightly biased in favour of females but marginally so and is probably accounted for in terms of greater female longevity. The ethnographic survey is nonetheless a true reflection of those who continuously dwell on the landscape.

One question of fundamental importance to the project required address: is it justified or possible for a non-indigene to adopt an African landscape paradigm in studies such as this? Can one evade the myopia of the 'Western Gaze' (see Bender 1999)? In other words can it ever be possible for an outsider, to understand or appreciate heterogeneous practices, symbols, processes, and meanings? Can one be detached from the colonial perspective? The Eurocentric perspective to African landscapes has been described as essentially an 'outside' one encapsulating notions of wilderness and backwardness and the promise of possession, protection and modernisation (Luig and von Oppen 1997: 20). This sits in opposition to the plethora of distinct local knowledges and interests. The postcolonial period has witnessed an intense process of creolisation between the imported and indigenous frames of reference. Featured elements cannot be essentialised into constituent elements. It is interesting to note that African languages have no equivalent term for either landscape or nature. Although it is clear the non-existence of such terms does not denote the non-conceptualisation of comparable abstractions (1997: 21). This notion of the syncretistic landscape perspective, the social history of African landscapes, and how this relates to power and perception has seen much focus (Beinart 1997; Darian-Smith *et al* 1996; Fairhead and Leach 1997; Harries 1997; Pratt 1992; Ranger 1997, 1999; Schumaker 1997). In response to the questions posed earlier it is asserted that one can reservedly posit justification for the non-indigene to research African landscapes. The historical process of making landscape undoubtedly differs in cross-cultural contexts. Nevertheless the ethnographer and the informers/subjects must become "active participants in an intercultural discourse" (Ucko and Layton 1999: 6) to further elucidate the material. There are of course benefits to be outside of the culture in question just as there are drawbacks. The outsider can better theorise difference and commonality and is in a better position to perceive subconscious cultural manifestations. On the other hand subtler areas of potential investigation might be potentially missed through lack of familiarity with subjects.

Communicative discourse (Eco 1990: 40-53) happens during face-to-face spoken interaction as in anthropological fieldwork. This is a technique for achieving intersubjective understanding across cultural boundaries. This is based on the phenomenology of Schutz. For Schutz intersubjectivity is the retrospective condition that facilitates our experience and meaning attribution of the world and the understanding that worldly significance is something we share with others. It thus needs to be remembered that "[t]he retrospective glance cast by indigenous peoples upon their own experience (including encounters with other anthropologists) draws upon events that occurred long before the arrival of the fieldworker" (Ucko 1999: 12). Thus ethnographic interview may make possible understanding across cultural and experiential boundaries. It is impossible to recover all the information required to interpret the invention of mythic and symbolic tradition (Moore 1970: 357). Nonetheless it is possible to highlight possibilities of circumstance that may have catalysed invention.

In addition to the formal structured interviews of the project various other data-gathering methods were employed including oral history, village meetings, conversational interviews, participatory observation and participatory appraisal, and field conversation. Interviews were conducted on an interviewee – interviewer/translator basis whenever possible. It was believed this would minimise the likelihood of repetition and leading or cueing statements. This interview style was not possible in all circumstances and at times group interviews had to be conducted. A standard interview guide is available below. Obviously questioning was not fixed to this template and frequently it cascaded into various subject areas as answers informed proceeding questions. There were also various periods of participant observation. Due to the cultural formalities associated with meeting and introduction the number of interviews conducted each day was somewhat less than originally expected. Initial interviews varied in length but usually lasted between 40-60 minutes. Subsequent interviews normally lasted for approximately 10-30 minutes. Each transect was surveyed in a period of approximately five weeks. There were also periods of secondary/archival research. This archival research was conducted at the Tanzanian National Archives, the University of Dar es Salaam Library, Kilimanjaro District Governmental Offices, and at the personal archives of Paramount Chief Thomas Marealle II in Moshi. In this sense the project adopted the *technique triangulation* method (Hammersley and Atkinson 1995: 230-2; see also Denzin 1989).

The project cross-referenced multiple lines of evidence to provide depth to the description of gathered social meanings. Procedural and interpretive inspiration was also gained from a methodological practice favoured by aid-agencies, government organisations, and resource management that has been labelled *Participatory Rural Appraisal* (see Chambers 1993). This low-cost method, in contrast to *Rapid Rural Appraisal*, incorporates informant bias as the key interpretive component (Tagseth 2000: 30). Tagseth's (2000: 30) comment that there is no substitute for prolonged and expensive fieldwork is an absolutely valid one. Nonetheless a triangulating form of Participatory Rural Appraisal provides an optimal package for researching in such settings being sensitive to both financial and methodological issues.

HOUSEHOLD INTERVIEW GUIDE

0) Introduction
   - 0.1. Introduction and details of field team
   - 0.2. Synopsis of research
   - 0.3. Anonymity issues and guarantees
   - 0.4. Exhibition of permits, contracts and letters of introduction

1) General Information
   - 1.1 What is your name?
   - 1.2 What is your position in the household?
   - 1.3 How old are you?
   - 1.4 What other persons belong to this household?
   - 1.5 What clan do you belong to?
   - 1.6 What religious denomination are you?
   - 1.7 What economic occupation/skills do you perform?

2) Wachagga and Clan Origins
   - 2.1 Are/Were your father and mother Wachagga?
   - 2.2 How many generations of your clan have lived on the mountain?
   - 2.3 Have these clan members always lived in this area?
   - 2.4 Where does your clan come from?
   - 2.4 For how many years have your clan lived on the mountain?
   - 2.5 Where do the Wachagga come from?
   - 2.6 Were any other people living on the mountain before the Wachagga?
   - 2.7 How and why are the Wachagga different from the other tribes in Northern Tanzania and Southern Kenya?
   - 2.8 Do you know anything about the tribal and clan wars?

3) Dwelling and Environment
   - 3.1 Why do the Wachagga live on the mountain?
   - 3.2 Why do you (and your family) live on the mountain?
   - 3.3 Does the mountain affect your daily activities?
   - 3.4 What crops do you cultivate?
   - 3.5 How much land is owned or rented by members of this household?
   - 3.6 How did you obtain your land?
   - 3.7 How many heads of cattle do you own?
   - 3.8 Where and how do you graze your cattle?
   - 3.9 Where do you get your irrigation and drinking water from?
   - 3.10 Does your household exploit any of the resources located in the forest and higher up the mountain?

4) Traditional Religion
   - 4.1 What religion was there before the missionaries?
   - 4.2 Do some people still practice this religion?
   - 4.3 Do you know of any pagans who live locally?
   - 4.4 What do the pagans believe and practice?
   - 4.5 Do you know or conduct any rituals that involve the mountain in any way? Please give examples of such practices and identify those who perform them.
   - 4.6 Why did people stop believing in the traditional religion and convert to Christianity (and other world religions)?

5) Sacred/Ritual Landscape
   - 5.1 Do you believe the mountain has special, supernatural or religious powers?

5.2 Do you know of person, clan or tribe that holds such beliefs about Mount Kilimanjaro?

5.3 Do you know of any features of the mountain or local landscape that you (or others) believe has special, supernatural or religious powers?

5.4 Do any places or features on the land cause things to happen? What and where are they and what sorts of things do they do?

5.5 Do you know anything about local waterfalls, rock formations, lakes, trees that makes them important?

5.6 Does the clan that this household belongs to have any special places? If so why are they special and what is done there?

5.7 Do you know of any local inscribed rocks, rock paintings or engravings that might have a religious or supernatural purpose?

5.8 Do certain people have special or religious knowledge of the mountain and its powers? If so who are they and what kinds of things do they know about?

6) Death and Afterlife

6.1 Do you hold a belief in an afterlife or life after death?

6.2 What do you know about what happens after a person dies?

6.3 Where do the dead go?

6.4 Has the afterlife got any relationship with the mountain?

6.5 Can the dead influence the lives of the living? If so how?

6.6 Is there ever any communication between the living and the dead?

SAMPLE PERMITS AND PERMISSIONS

TANZANIA COMMISSION FOR SCIENCE AND TECHNOLOGY
(COSTECH)

Telegrams: COSTECH
Telephones: (255 - 22) 2700745-6
Director General: (255 - 22) 2700750
Fax: (255 - 22) 2775313
Telex: 41177 UTAFITI
E-M: Rclearance@costech.or.tz

Ali Hassan Mwinyi Road
Kijitonyama Area
P.O. Box 4302
Dar es Salaam
Tanzania

RESEARCH PERMIT

No. 2004 – 139 - NA- 2004 – 12

Date: 14th July 2004

1. Name : **Timothy A. Clack**

2. Nationality : **British**

3. Title : **Archaeological Investigation into the Origin (Myhs) of the Chagga of Kilimanjaro**

4. Research shall be confined to the following region(s): **Kilimanjaro**

5. Permit validity: **14th July 2004 to 13th July 2005**

6. Local Contact/Collaborator: **Bertram Mapunda, Archaeological Unit, University of Dar es Salaam**

7. Researcher is required to submit progress report on quarterly basis and submit all Publications made after research.

H.P. Gideon
for: DIRECTOR GENERAL

JAMHURI YA MUUNGANO WA TANZANIA
OFISI YA RAIS
TAWALA ZA MIKOA NA SERIKALI ZA MITAA

MKOA WA KILIMANJARO
Anwani ya Simu: 'REGCOM' KILIMANJARO
Simu Na. Moshi 027-2754236/7
Fax Na. 027-2752164
Unapojibu tafadhali taja:

OFISI YA MKUU WA MKOA ,
S.L.P. 3070,
MOSHI.

27 Julai, 2004

Kumb. Na.E.10/25/191

Makatibu Tawala wa Wilaya,
Same, Mwanga, Hai,
Moshi na Rombo.

Yah: **MAOMBI YA KIBALI CHA KUFANYA UTAFITI**
TIMOTHY A. CLACK

Tafadhali husika na kichwa cha habari hapo juu.

Bw. Timothy A. Clack anatambulishwa kwetu na Tume ya Sayansi na Teknlolojia (COSTECH) na kuombewa kibali cha kufanya utafiti hapa mkoani kwetu.

Ikiwa mtaridhika na anachohitaji kuhusiana na utafiti huo mruhusuni afanye utafiti.

O.B. Msuya
Kny: KATIBU TAWALA MKOA
KILIMANJARO

89

TANZANIA COMMISSION FOR SCIENCE AND TECHNOLOGY
(COSTECH)

Telegrams: COSTECH
Telephones: (255 - 22) 2775155 - 6, 2700745-6
Director General: (255 - 22) 2700750
Fax: (255 - 22) 2775313
Telex: 41177 UTAFITI
E-M: Rclearance@costech.or.tz

Ali Hassan Mwinyi Road
Kijitonyama Area
P.O. Box 4302
Dar es Salaam
Tanzania

In reply please quote: **CST/RCA 2004/12/1582/2004**   **Date:** 14th July 2004

Director of Immigration Services
Ministry of Home Affairs
P.O. Box 512
DAR ES SALAAM

Dear Sir/Madam,

RESEARCH PERMIT

We wish to introduce to you **Timothy A. Clack from UK** who has been granted a research permit No. **2004 – 139 - NA- 2004 – 12** dated 14th **July 2004.**

The permit allows him/her to do research in the country entitled **Archaeological Investigation into the Origin (Myhs) of the Chagga of Kilimanjaro.**

We would like to support the application of the researcher(s) for the appropriate immigration status to enable the scholar(s) begin research as soon as possible.

By copy of this letter, we are requesting regional authorities and other relevant institution status to accord the researcher(s) all the necessary assistance. Similarly the designated local contact is requested to assist the researcher(s)

Yours faithfully,

H.P. Gideon
for: **DIRECTOR GENERAL**
cc:
1. Regional Administrative Secretary: **Kilimanjaro**

2. Local Advisor: **Bertram Mapunda, Archaeological Unit, University of Dar es Salaam**

3. Co-researchers: **None**

OFISI YA RAIS
TAWALA ZA MIKOA NA SERIKALI ZA MITAA
HALMASHAURI YA WILAYA YA MOSHI
(Barua zote ziandikwe kwa Mkurugenzi Mtendaji)

S.L.P. 6924,
Moshi

Mkoa wa Kilimanjaro
Simu 2755172/2751865 Fax: 2752336

Kumb Na. E.10/16/Vol.II/130.   27/07/2004

Maafisa Watendaji wa Kata,
Marangu Mashariki/Magharibi.

YAH: KUFANYA UTAFITI TIMOTHY A. CLACK

Mtajwa hapo juu ameruhusiwa kufanya utafiti katika Wilaya ya Moshi juu ya Archaeological Investigation into the Origin (Myhs) of the Chagga of Kilimanjaro.

Utafiti unaanza tarehe 14th July 2004 13th July 2005. Tafadhali mpatie msaada atakaohitaji.

Kwa nakala ya barua hii Maafisa Wtendaji wa kata wote Halmashauri ya Wilaya Moshi wanaombwa kumpa msaada kama atahitaji toka kata zao.

Robert M. Kitimbo,
KAIMU MKURUGENZI MTENDAJI
HALMASHAURI YA WILAYA
MOSHI.

Mkurugenzi Mtendaji
Halmashauri ya Wilaya
Moshi

Nakala kwa:-

Maafisa Watendaji Wote,
Halmashauri ya Wilaya,
Moshi.

Ndugu Timothy A. Clack.

JAMHURI YA MUUNGANO WA TANZANIA
OFISI YA RAIS
TAWALA ZA MIKOA NA SERIKALI-ZA MITAA

MKOA WA KILIMANJARO
Simu ya upepo: ADMIN MOSHI
Simu:027-2752211
Fax: 027-2752184
E-Mail: raskilimanjaro@yahoo.co.uk

OFISI YA MKUU WA WILAYA,
S.L.P. 3042,
MOSHI.

Kwa Kujibu Taja
Kumb.Na: E.10/29/IV/37

27 Julai, 2004

Mkurugenzi Mtendaji,

Halmashauri ya Wilaya,

MOSHI.

YAH: KUFANYA UTAFITI - TIMOTHY A. CLACK

    Mtajwa hapo juu anatambulishwa kwetu na Tume ya Sayansi na
Teknolojia (COSTECH) kufanya Utafiti katika wilaya yetu ya Moshi juu
ya "Archaeological Investigation into the Origin (Myhs) of the Chagga
of Kilimanjaro.

    Utafiti wake unaanza tarehe 14 Julai, 2004 hadi tarehe
13 Julai, 2005. Wakati akiwa katika eneo lako unaombwa kumpatia
kila aina ya ushirikiano ili apate kufanikisha utafiti wake.

M. J. Mwanga
KATIBU TAWALA WILAYA
MOSHI

KATIBU TAWALA WILAYA
MOSHI

Nakala kwa:- Makatibu Tarafa,
        Hai Mashariki, Kibosho,
        Vunjo Mashariki na Magharibi.

    "   "   Mr. Timothy A. Clack.

JAMHURI YA MUUNGANO WA TANZANIA
OFISI YA RAIS
TAWALA ZA MIKOA NA SERIKALI ZA MITAA

WILAYA YA ROMBO
Simu Nambari
2757133/2757106
Fax No 027-2757133
Unapojibu tafadhali taja:

OFISI YA MKUU WA WILAYA
S.L.P. 2
Mkuu Rombo

Kumb. Na T.40/18/197

9/8/ 2004

KWA YEYOTE ANAYEHUSIKA
ROMBO.

YAH;   UTAFITI

Tafadhali husika na habari niliyoitaja hapo juu.

Namtambulisha kwako Ndugu **TIMOTHY A. CLARK** amabaye amepewa
kibali cha kufanya utafiti katika Wilaya yetu ya Rombo juu ya **ARCHAEOLOGIOAL**
INVESTIGATION INTO THE ORIGIN (MYTHS) OF THE CHAGGA OF
KILIMANJARO kuanzia tarehe 9/8/2004 hadi 13/8/2004.

Tafadhali msikilize na msaada atakaouhitaji kutoka kwako

A. D. SHIRIMA
Kny; **KATIBU TAWALA WILAYA**
ROMBO
kny KATIBU TAWALA WA WILAYA
ROMBO

Nakala kwa : **KATIBU TARAFA**
MKUU

REFERENCES

PUBLISHED SOURCES

ABBINK, J. 1998. An Historical-Anthropological Approach to Islam in Ethiopia: Issues of Identity and Politics. *Journal of African Cultural Studies* 11(2): 109-24.

ABRAHAMS, P. W. and J. A. PARSONS 1996. Geophagy in the Tropics: A Literature Review. *Geographical Journal* 162: 63-72.

ABUNGU, G. H. O. 1994. "Islam on the Kenyan Coast: An Overview of Kenyan Coastal Sacred Sites," in D. L. Carmichael, J. Hubert, B. Reeves and A. Schanche (eds) *Sacred Sites, Sacred Places.* pp. 152-62. London: Routledge.

ALTHAUS, G. 1992. *Mamba: Anfang in Afrika.* Erlangen: Verlag der Evangelisch-Lutherisches Mission.

ANDREA, S. W. and S. H. KATZ 1998. Geophagy in Pregnancy: A Test of a Hypothesis. *Current Anthropology* 39(4): 532-45.

ANDRESEN, J. 2001. "Conclusion: Religion in the Flesh: Forging New Methodologies for the Study of Religion," in J. Andresen, (ed.) *Religion in Mind: Cognitive Perspectives on Religious Belief, Ritual, and Experience.* pp. 257-87. Cambridge: Cambridge University Press.

ARGYLE, M. 2000. *Psychology and Religion.* London: Routledge.

BAILEY, P. M. J. 1968. The Changing Economy of the Chagga of Marangu, Kilimanjaro. *Geography* 53: 163-9.

BEINART, W. 1997. Vets, Viruses and Environmentalism: The Cape in the 1870s and 1880s (South Africa). *Paideuma* 43: 227-51.

BEINART, W. and J. McGREGOR (eds) 2003. *Social History and African Environments.* London: James Currey.

BELL, T. and G. LOCK 2000. "Topographic and Cultural Influences on Walking the Ridgeway in Later Prehistoric Times," in G. Lock (ed.) *Beyond the Map: Archaeology and Spatial Technologies.* pp. 85-100. Amsterdam: Lightning Source.

BENDER, B. 1992. Theorising Landscapes, and the Prehistoric Landscape of Stonehenge. *Man* 27: 735-55.

BENDER, B. 1993. "Introduction – Landscape – Meaning and Action," in B. Bender (ed.) *Landscape: Politics and Perspectives.* pp. 1-17. Oxford: Berg.

BENDER, B. 1999. "Subverting the Western Gaze: Mapping Alternative Worlds," in P. J. Ucko and R. Layton (eds) *The Archaeology and Anthropology of Landscape: Shaping Your Landscape.* pp. 31-45. London: Routledge.

BENDER, B., S. HAMILTON and C. TILLEY 1997. Leskernick: Stone Worlds; Alternative Narratives; Nested Landscapes. *Proceedings of the Prehistoric Society* 63: 147-78.

BENNETT, N. R. 1964. The British on Kilimanjaro 1884-1892. *Tanzania Notes and Records* 63: 229-44.

BENSON, J. S. 1974. *A Study of the Religious Beliefs of the Maasai Tribe and Implications on the Work of the Evangelical Lutheran Church in Tanzania.* Unpublished Master's thesis, Sacred Theology of the Northwestern Lutheran Seminary.

BENTON, T. 1994. "Biology and Social Theory in the Environmental Debate," in M. Redclift and T. Benton (eds) *Social Theory and the Global Environment.* pp. 28-50. London: Routledge.

BIERLICH, B. 1999. Sacrifice, Plants, and Western Pharmaceuticals: Money and Health Care in Northern Ghana. *Medical Anthropology Quarterly* 13(3): 316-37.

BREWIN, D. R. 1965. Kilimanjaro Agriculture. *Tanganyika Notes and Records* 64: 115-7.

BROPHY, K. 1999. Seeing the Curcus as a Symbolic River. *British Archaeology* 44: 6-7.

BRÜCK, J. 1998. In the Footsteps of the Ancestors: A Review of Tilley's *A Phenomenology of Landscape, Places, Paths and Monuments. Archaeological Review from Cambridge* 15: 23-36.

BRÜCK, J. 2001a. Monuments, Power and Personhood in the British Neolithic. *Journal of the Royal Anthropological Institute* 7: 649-67.

BRÜCK, J. 2001b. "Beyond Metaphors and Technologies of Transformation in the English Middle and Late Bronze Age," in J. Brück (ed.) *Bronze Age Landscapes: Tradition and Transformation.* pp. 65-82. Oxford: Oxbow.

BRÜCK, J. 2005. Experiencing the Past? The Development of a Phenomenological Archaeology in British Prehistory. *Archaeological Dialogues* 12(1): 45-72.

BUNZL, M. 1996. "Franz Boas and the Humboldtian Tradition: From Volksgeist and Nationalcharakter to an Anthropological Concept of Culture," in G. W. Stocking (ed.) *Volksgeist as Method and Ethic: Essays on Boasian Ethnography and the German Anthropological Tradition.* pp. 17-78. Madison: University of Wisconsin Press.

BUSBY, C. 1997. Permeable and Partible Persons: A Comparative Analysis of Gender and the Body in South India and Melanesia. *Journal of the Royal Anthropological Institute* 3: 261-78.

CAMPBELL, J. 1999. Nationalism, Ethnicity and Religion: Fundamental Conflicts and the Politics of Identity in Tanzania. *Nations and Nationalism* 5(1): 105-2.

CARMICHAEL, D. L., J. HUBERT and B. REEVES 1994. "Introduction," in D. L. Carmichael, J. Hubert, B. Reeves and A. Schanche (eds) *Sacred Sites, Sacred Places.* pp. 1-8. London: Routledge.

CENTRAL CENSUS OFFICE. 2003. *2002 Population and Housing Census General Report.* Dar es Salaam: Government Printers.

CHADWICK, A. 2004. "Geographies of Sentience: An Introduction to Space, Place and Time," in A. Chadwick (ed.) *Stories from the Landscape: Archaeologies of Inhabitation.* pp. 1-31. Oxford: BAR Publishing.

CHAGGA COUNCIL 1955. *Recent Trends in Chagga Political Development.* Moshi: Kilimanjaro Native Co-operative Union Printing Press.

CHAMBERS, R. 1993. *Challenging the Professions: Frontiers for Rural Development.* London: Intermediate Technology Publications.

CHAPMAN, H. P. and B. R. GEAREY 2000. Palaeoecology and the Perception of Prehistoric Landscapes: Some Comments on Visual Approaches to Phenomenology. *Antiquity* 74: 316.

CHAMBERS, I. 1986. *Popular Culture: The Metropolitan Experience.* London: Methuen.

CHARSLEY, S. 1992. "Dreams in African Churches." in M. C. Jedrej and R. Shaw (eds) *Dreaming, Religion and Society in Africa.* pp 153-76. New York: E. J. Brill.

CLACK, T. A. R. 2004. "Neurophenomenology: Worthwhile Research Direction for the Archaeological Study of Religion." in T. Insoll (ed.) *Belief in the Past: The Proceedings of the 2002 Manchester Conference on Archaeology and Religion.* pp. 51-61. Oxford: BAR Publishing.

CLACK, T. A. R. 2005. Protective Memoryscapes of the Chagga of Kilimanjaro, Tanzania. *Azania* 40: 110-7.

CLIMO, J. J. and M. G. CATTELL (eds) 2002a. *Social Memory and History: Anthropological Perspectives.* Walnut Creek, CA: Altamira Press.

CLIMO, J. J. and M. G. CATTELL 2002b. "Preface," in J. J. Climo and M. G. Cattell (eds) *Social Memory and History: Anthropological Perspectives.* pp. ix-xi. Walnut Creek, CA: Altamira Press.

CLIMO, J. J. and M. G. CATTELL 2002c. "Introduction: Meaning in Social Memory and History: Anthropological Perspectives," in J. J. Climo and M. G. Cattell (eds) *Social Memory and History: Anthropological Perspectives.* pp. 1-36. Walnut Creek, CA: Altamira Press.

COCHETTI, S. 1995. The Dogon Sacrifice as a Literal Metaphor. *Paragrana: Internationale Zeitschrift fur Historische Anthropologie* 4(2): 144-50.

COHEN, D. W. and E. S. A. ODHIABO 1989. *Siaya: The Historical Anthropology of an African Landscape.* Nairobi: Heinemann.

COLSON, E. 1997. Places of Power and Shrines of the Land. *Paideuma* 43: 47-57.

COLWELL, A. S. C. 2000a. Fertility and Mortality Trends in Colonial Kilimanjaro. Working Paper. RP174.
http://www.hsph.harvard.edu/takemi/rp174.pdf, accessed January 2005.

COLWELL, A. S. C. 2000b. *Vision and Revision: Demography, Material and Child Health Development, and the Representation of Women in*

Colonial Tanzania. Unpublished PhD thesis, University of Illinois,

COMAROFF, J. 1985. *Body of Power, Spirit of Resistance: The Culture and History of a South African People.* Chicago: University of Chicago Press.

COMAROFF, J. and J. COMAROFF 1991. *Of Revelation and Revolution: Christianity, Colonialism, and Consciousness in South Africa.* Chicago: University of Chicago Press.

COMAROFF, J. and J. COMAROFF 1992. *Ethnography and the Historical Imagination.* Boulder: Westview.

COMAROFF, J. and J. COMAROFF 1997. *Of Revelation and Revolution: The Dialectics of Modernity on a South African Frontier.* Chicago: University of Chicago Press.

CONYERS, D., F. KASSULEMEMBA, A. MOSHA and P. NNKO 1970. *Agro-Economic Zones of North-Eastern Tanzania.* Dar es Salaam: Bureau of Resource Assessment and Land-Use Planning.

CORR, R. 2003. Ritual, Knowledge, and the Politics of Identity in Andean Festivities. *Ethnology.* 42(1): 39-54.

COSGROVE, D. 1989. "Geography is Everywhere: Culture and Symbolism in Human Landscapes," in D. Gregory and R. Walford (eds), *Horizons in Human Geography.* London: Macmillan.

COSGROVE, D. 1993. "Landscapes and Myths, Gods and Humans," in B. Bender (ed.) *Landscape: Politics and Perspectives.* pp. 281-305. Oxford: Berg.

CSORDAS, T. J. 1994. "Introduction: The Body as Representation and Being-in-the-World." in T. J. Csordas (ed.) *Embodiment and Experience: The Existential Ground of Culture and Self.* pp. 1-24. Cambridge: Cambridge University Press.

CSORDAS, T. J. 1999. "Embodiment and Cultural Phenomenology," in G. Weiss and H. Fern Haber (eds) *Perspectives on Embodiment: The Intersections of Nature and Culture.* pp. 143-62. London: Routledge.

CUMMINGS, V. 2000. "Landscapes in Motion: Interactive Computer Imagery and Neolithic Landscapes of the Outer Hebrides," in C. Buck, V. Cummings, C. Henley, S. Mills and S. Trick (eds) *UK Chapter of Computer Applications and Quantitative Methods in Archaeology.* pp. 11-20. Oxford: Achaeopress.

CUMMINGS, V. 2002a. Between the Mountains and the Sea: A Reconsideration of the Neolithic Monuments of South-West Scotland. *Proceedings of the Prehistoric Society* 68: 125-46.

CUMMINGS, V. 2002b. Experiencing Texture and Transformation in the British Neolithic. *Oxford Journal of Archaeology* 21: 249-61.

CUMMINGS, V., A. JONES and A. WATSON 2002. Divided Places: Phenomenology and Asymmetry in the Monuments of the Black Mountains, Southeast Wales. *Cambridge Archaeological Journal* 12: 57-70.

CUMMINGS, V. and A. WHITTLE 2003. Tombs with a View: Landscape, Monuments and Trees. *Antiquity* 77: 255-66.

CUMMINS, D. D. 1998. "Social Norms and Other Minds: The Evolutionary Roots of Higher Cognition," in D. D. Cummins and C. Allen (eds) *The Evolution of Mind.* pp. 30-50. Oxford: Oxford University Press.

CUPITT, D. 1984. *The Sea of Faith: Christianity and Change.* London: Collins.

CURRY, J. R. 1939. Eleusine Cultivation by the Wachagga on Kilimanjaro. *East African Agricultural Journal* 5: 386-90.

DALFOVO, A. T. 1997. The Lugbara Ancestors. *Anthropos* 92(4): 485-500.

DANIELSON, E. R. 1977. *Forty Years with Christ in Tanzania, 1928-1968.* New York: Lutheran Church in America.

DARIAN-SMITH, K., L. GUNNER and S. NUTTALL (eds) 1996. *Text, Theory and Space: Land, Literature and History in South Africa and Australia.* London: Routledge.

DARVILL, T. 1999. "The Historic Environment, Historic Landscapes, and Space-Time-Action Models in Landscape Archaeology," in P. J. Ucko and R. Layton (eds) *The Archaeology and Anthropology of Landscape: Shaping Your Landscape.* pp. 104-18. London: Routledge.

DAVENPORT, G. 1984. *The Geography of Imagination.* London: Picador.

DENZIN, N. K. 1989. *The Research Act: A Theoretical Introduction to Sociological Methods.* Englewood Cliffs, NJ: Prentice Hall.

DEVEREUX, S. and J. HODDINOTT 1992. "Issues in Data Collection," in S. Devereux and J. Hoddinott (eds) *Fieldwork in Developing Countries.* pp. 25-40. Boulder, CL: Lynne Rienner.

DOVEY, K. 2000. "The Quest for Authenticity and the Replication of Environmental Meaning," in D. Seamon and R. Mugerauer (eds) 2000. *Dwelling, Place and Environment: Towards a Phenomenology of Person and World.* pp. 33-49. Malabar, FL: Krieger.

DREYFUS, H. L. 1972. *What Computers Can't Do: A Critique of Artificial Intelligence.* New York: Harper & Row.

DREYFUS, H. L. 1992. *What Computers Still Can't Do: A Critique of Artificial Reason.* Cambridge, MA: MIT Press.

DUDLEY, G. 1977. *Religion on Trial: Mircea Eliade and His Critics.* Philadelphia: Temple University Press.

DUNDAS, C. 1968. *Kilimanjaro and Its People: A History of the Wachagga, their Laws, Customs and Legends, together with Some Account of the Highest Mountain in Africa.* London: Frank Cass & Co.

ECO, U. 1990. *The Limits of Interpretation.* Bloomington: Indiana University Press.

EDMONDS, M. R. 1999. *Ancestral Geographies of the Neolithic.* London: Routledge.

EDMONDS, M. and T. G, McELEARNEY 1999. Inhabitation and Access: Landscapes and the Internet at Gardom's Edge. *Internet Archaeology* 6, http://intarch.ac.uk/journal/issue6/edmonds_index.html (last accessed 15th October 2005).

ELIADE, M. 1957. *The Sacred and the Profane: The Nature of Religion.* W. R. Trask (trans.). London: Harcourt.

ELIADE, M. 1958. *Patterns in Comparative Religion.* R. Sheed (trans.). London: University of Nebraska Press.

ELIADE, M. 1960. *Myths, Dreams and Mysteries.* New York: Harper and Brothers.

ELIADE, M. 1963. *Myth and Reality.* W. R. Trask (trans.). New York: Harper & Row.

ELIADE, M. 1969. *The Quest.* Chicago: University of Chicago Press.

ELIADE, M. 1973. *Australian Religions: An Introduction.* Ithaca: Cornell University Press.

ELLUL, J. 1963. "The Technological Order." in C. Stover (ed.) *The Technological Order.* Detroit: Wayne State University Press.

EMANATIAN, M. 1996. Everyday Metaphors of Lust and Sex in Chagga. *Ethos* 24(2): 195-236.

EVANGELICAL LUTHERAN CHURCH IN TANZANIA 1909. *Kitabu kya Fiimbo na Katekisimo ya Kimashami* (Hymn Book and Pentateuch Text in Kimashami). Moshi: Moshi Lutheran Press.

FAIRHEAD, J. and M. LEACH 1997. Deforestation in Question: Dialogue and Dissonance in Ecological, Social and Historical Knowledge. *Paideuma* 43: 193-225.

FEIERMAN, S. 1990. *Peasant Intellectuals: Anthropology and History in Tanzania.* Madison: University of Wisconsin Press.

FELD, S. and K. H. BASSO (eds) 1996. *Senses of Place.* Santa Fe: School of American Research.

FERNANDES, E. C. M., A. O'KTINGATI, and J. MAGHEMBE 1984. The Chagga Homegarden: A Multistoried Agroforestry Cropping System on Mt Kilimanjaro (Northern Tanzania). *Agro-Forestry Systems* 2: 73-86.

FERNANDES, E. C. M. and P. K. R. NAIR 1986. An Evaluation of the Structure and Function of Tropical Homegardens. *Agricultural Systems* 21: 279-310.

FIELDER, K. 1983. *Christentum und Afrikanische Kultur: Konservative Deutsche Missionare in Tanzania, 1900-1940.* Gütersloh: Gerd Mohn.

FIELDER, K. 1996. Christianity and African Culture: Conservative German Protestant Missionaries in Tanzania 1900-1940. Leiden: E. J. Brill.

FISIY, C. P. and P. GESCHIERE 1996. "Witchcraft, Violence and Identity: Different Trajectories in Postcolonial Cameroon," in R. Werbner and T. Ranger (eds) *Postcolonial Identities in Africa.* pp. 193-221. London: Zed.

FLATT, D. C. 1980. *Man and Deity in an African Society: A Study of Religious Meaning and Value among the Ilarusa of Northern Tanzania.* Dubuque: Lutheran Church in America.

FLEMING, A. 1999. Phenomenology and the Megaliths of Wales: A Dreaming Too Far. *Oxford Journal of Archaeology* 18: 119-25.

FOSBROKE, H. A. 1935. The Defensive Measures of Certain Tribes in North-Eastern Tanganyika. *Tanganyika Notes and Records* 35: 1-6.

FOSBROKE, H. A. 1954. Chagga Forts and Bolt Holes. *Tanganyika Notes and Records* 37: 115-29.

FOSBROKE, H. A. and P. I. Marealle 1952. *The Engraved Rocks of Kilimanjaro. Man* 52: 179-81.

FOSBROKE, H. A. and H. SASSOON 1965. Archaeological Remains on Kilimanjaro. *Tanganyika Notes and Records* 64: 62-3.

FOWLER, C. 2002. "Body Parts: Personhood and Materiality in the Earlier Manx Neolithic," in Y. Hamilakis, M. Pluciennik and S. Tarlow (eds) *Thinking Through the Body: Archaeologies of Corporeality.* pp. 47-69. Amsterdam: Kluwer.

FOWLER, C. 2004. *The Archaeology of Personhood: An Anthropological Approach.* London: Routledge.

FRASER, S. 1998. The Public Forum and the Space Between: The Materiality of Social Strategy in the Irish Neolithic. *Proceedings of the Prehistoric Society* 64: 204-24.

GADAMER, H-G. 1976. *Philosophical Hermeneutics.* Berkeley, CA: University of California Press.

GARDINER, J. M. and R. I. JAVA 1993. "Recognising and Remembering," in A. F. Collins, S. E. Gathercole, M. A. Conway and P. E. Morris (eds) *Theories of Memory.* pp. 163-88. Hove: Erlbaum.

GATWA, T. 1999. Victims or Guilty? Can the Rwandan Churches Repent and Bear the Burden of the Nation for the 1994 Tragedy? *International Review of Mission* 88: 347-63.

GELL, A. 1985. How to Read a Map: Remarks on the Practical Logic of Navigation. *Man* 20: 271-86.

GIDDENS, A. 1979. *Central Problems in Social Theory.* London: Macmillan.

GIDDENS, A. 1984. *The Constitution of Society: Outline of the Theory of Structuration.* Cambridge: Polity Press.

GILL, S. 1998. *Storytracking.* Oxford: Oxford University Press.

GILLINGHAM, M. E. 1997. *Gaining Access to Water: Indigenous Irrigation on Mount Kilimanjaro,* *Tanzania.* Unpublished PhD thesis, University of Cambridge, Cambridge.

GILLINGHAM, M. E. 1999. Gaining Access to Water: Formal and Working Rules of Indigenous Irrigation Management on Mount Kilimanjaro, Tanzania. *Natural Resources Journal* 39(3): 419-41.

GOLLEDGE, R. G. 2003. "Human Wayfinding and Cognitive Maps," in M. Rockman and J. Steele (eds) *Colonization of Unfamiliar Landscapes: The Archaeology of Adaptation.* pp. 25-43. London: Routledge.

GOSDEN, C. 1994. *Social Being and Time.* Oxford: Blackwell.

GOSDEN, C. 1996. Can we take the Aryan out of Heideggerian? *Archaeological Dialogues* 3: 22-5.

GOSDEN, C. and G. LOCK 1998. Prehistoric Histories. *World Archaeology* 30(1): 2-12.

GOW, P. 1994. "Against the Motion (2)", in J. Weiner (ed.), *Aesthetics in a Cross-Cultural Category.* Manchester: GDAT.

GOW, P. 1995. "Land, People, and Paper in Western Amazonia", in E. Hirsch and M. O'Hanlon (eds), *The Anthropology of Landscape.* Oxford: Clarendon Press.

GREENE, S. E. 1997. Sacred Terrain: Religion, Politics and Place in the History of Anloga (Ghana). *International Journal of African Historical Studies* 30(1): 1-22.

GREENFIELD, P. J. 1996. Self, Family, and Community in White Mountain Apache Society. *Ethos* 24(3): 491-509.

GREENFIELD, S. M. (ed) 2001. *Reinventing Religions: Syncretism and Transformation in Africa and the Americas.* Lanham: Rowman & Littlefield.

GROVE, A. 1993. Water Use by the Chagga on Kilimanjaro. *African Affairs* 92(368): 431-48.

GUTHRIE, S. 1993. *Faces in the Clouds: A New Theory of Religion.* Oxford: Oxford University Press.

GUTHRIE, S. 2001. Comment: Rethinking Animism. *Journal of the Royal Anthropological Institute* 7(1): 156-7.

GUTMANN, B. 1909. *Dichten und Denken der Dschagganeger.* Leipzig.

GUTMANN, B. 1926. *Das Recht der Dschagga.* Munich: C. H. Beck.

GUTMANN, B. and E. JESSEN 1905. Nachrichten aus Madschame. *Evangelisch-Lutherisches Missionsblatt* 60(20): 490-3.

HÅKANSSON, N. T. 1995. Irrigation, Population Pressure, and Exchange in Precolonial Pare, Tanzania. *Research in Economic Anthropology* 16: 297-323.

HÅKANSSON, N. T. 1998. Rulers and Rainmakers in Precolonial South Pare, Tanzania: Exchange and Ritual Experts in Political Centralization. *Ethnology* 37: 263-83.

HAMILAKIS, Y. 2002. "The Past as Oral History: Towards an Archaeology of the Senses," in Y. Hamilakis, M. Pluciennik and S. Tarlow (eds) *Thinking through the Body: Archaeologies of Corporeality.* pp. 121-36. London: Kluwer.

HAMMERSLEY, M. and P. ATKINSON 1995. *Ethnography: Principles in Practice.* London: Routledge.

HARRIES, P. 1997. Under Alpine Eyes: Constructing Landscape and Society in Late Pre-Colonial South-East Africa. *Paideuma* 43: 171-92.

HARRIS, G. C. 1978. *Casting Out Anger: Religion among the Taita of Kenya.* Cambridge: Cambridge University Press.

HARRIS, W. T. 1950. "The Idea of God Among the Mende", in E. W. Smith (ed.) *African Ideas of God.* pp.277-99. London: Edinburgh House Press.

HARVEY, G. H. 2005. *Animism: Respecting the Living World.* London: Hurst & Company.

HASU, P. 1999. *Death and Desire: History through Ritual Practice in Kilimanjaro.* Saarijarvi: Gummerus Kirjapaino Oy.

HEIDEGGER, M. 1962. *Being and Time.* trans. J. Macquarrie and E. Robinson. Oxford: Blackwell.

HEIDEGGER, M. 1969. *Identity and Difference.* J. Stambaugh (ed.). New York: Harper & Row.

HEIDEGGER, M. 1971a. *Poetry, Language, Thought.* trans. A. Hofstadter. New York: Harper and Row.

HEIDEGGER, M. 1971b. *On the Way to Language.* trans. P. D. Hertz. New York: Harper Row.

HEIDEGGER, M. 1977. *The Question Concerning Technology and Other Essays.* trans. W. Lovitt. New York: Harper and Row.

HEIDEGGER, M. 1992. *Parmenides.* A. Schuwer and R. Rojcewicz (trans). Bloomington, IN: Indiana University Press.

HEIDEGGER, M. 1993. "Building Dwelling Thinking." in D. F. Krell (ed.) *Martin Heidegger: Basic Writings.* pp. 41-87. London: Routledge.

HEIDEGGER, M. 1993b. "The Origin of the Work of Art," in D. F. Krell (ed.) *Martin Heidegger: Basic Writings.* London: Routledge.

HEMINGWAY, E. M. 2004 [1939]. *The Snows of Kilimanjaro: And Other Stories.* London: Vintage.

HENDRICKSON, H. 1996. "Introduction," in H. Hendrickson (ed.) *Clothing and Difference.* pp. 1-16. Durham: Duke University Press.

HERMANSEN, J. E., F. BENEDICT, T. CORNELIUSSEN, T. HOFSTEN and H. VENVIK 1985. *Catchment Forestry in Tanzania: Status and Management.* Oslo: Institute for Environmental Analysis.

HIND, D. 2004. "Where Many Paths Meet: Toward an Integrated Theory of Landscape and Technology," in A. M. Chadwick (ed.) *Stories from the Landscape: Archaeologies of Inhabitation.* pp. 35-51. Oxford: BAR Publishing.

HINDE, R. A. 1999. *Why Gods Persist: A Scientific Approach to Religion.* London: Routledge.

HINNEBUSCH, T. and D. NURSE 1981. Spirantization in Chaga. *SUGIA* 3: 51-78.

HIRSCH, E. 1995. "Introduction: Landscape: Between Place and Space", in E. Hirsch and M. O'Hanlon (eds), *The Anthropology of Landscape.* Oxford: Clarendon Press.

HIRSCH, E. S. (ed.) 1996. *The Block Book.* Washington, DC: National Association for the Education of Young Children.

HOBSBAWM, E. and T. RANGER 1983. *The Invention of Tradition.* Cambridge: Cambridge University Press.

HODDER, I. 1986. *Reading the Past: Current Approaches to Interpretation in Archaeology.* Cambridge: Cambridge University Press.

HODDER, I. 1999. *The Archaeological Process: An Introduction.* Oxford: Blackwell.

HOJBJERG, C. K. 2002. Inner Iconoclasm: Forms of Reflexivity in Loma Rituals of Sacrifice. *Social Anthropology* 10(1): 57-75.

HOLAND, I. S. 1996. *More People, More Trees: Population Growth, the Chagga Irrigation System, and the Expansion of a Sustainable Agroforestry System on Mount Kilimanjaro.* Unpublished PhD thesis, Norwegian University of Science and Technology, Trondheim.

HORNER, R., C. LACKEY, K. KOLASA, and K. WARREN 1991. Pica Practices among Pregnant Women. *Journal of the Dietetic Association* 1: 34-8.

HORTON, R. 1971. African Conversion. *Africa* 41(2): 85-108.

HORTON, R. 1993. *Patterns of Thought in Africa and the West: Essays on Magic, Religion and Science.* Cambridge: Cambridge University Press.

HOWARD, M. and A. MILLIARD (eds) 1997. *Hunger and Shame: Child Malnutrition and Poverty of Mount Kilimanjaro.* London: Routledge.

HUBERT, J. 1994."Sacred Beliefs and Beliefs of Sacredness," in D. L. Carmichael, J. Hubert, B. Reeves and A. Schanche (eds) *Sacred Sites, Sacred Places.* pp. 9-19. London: Routledge.

HUCHZERMEYER, F. W. 2003. *Crocodiles: Biology, Husbandry and Diseases.* Oxford: Oxford University Press.

HUNTINGTON, R. and P. METCALF 1979. *Celebrations of Death: The Anthropology of Mortuary Ritual.* Cambridge: Cambridge University Press.

ILIFFE, J. 1979. *A Modern History of Tanganyika.* Cambridge: Cambridge University Press.

INGOLD, T. 1992. Foraging for Data, Camping with Theories: Hunter-gatherers and Nomadic Pastoralists in Archaeology and Anthropology. *Antiquity* 66: 790-803.

INGOLD, T. 1993. "The Temporality of the Landscape". *World Archaeology* 25(2): 152-74.

INGOLD, T. 1995. "Building, Dwelling, Living: How Animals and People Make Themselves at Home in the World." in M. Strathern (ed.) *Shifting Contexts: Transformations in Anthropological Knowledge.* pp. 57-80. London: Routledge.

INGOLD, T. 2000. *The Perception of the Environment: Essays in Livelihood, Dwelling and Skill.* London: Routledge.

INSOLL, T. 2004a. *Archaeology, Ritual, Religion.* London: Routledge.

INSOLL, T. (ed.) 2004b. *Belief in the Past: The Proceedings of the 2002 Manchester Conference on Archaeology and Religion.* Oxford: BAR Publishing.

INSOLL, T. 2004c. "Are Archaeologists Afraid of Gods? Some Thoughts on Archaeology and Religion," in T. Insoll (ed.) *Belief in the Past: The Proceedings of the 2002 Manchester Conference on Archaeology and Religion.* pp. 1-6. Oxford: BAR Publishing.

JACKSON, P. 1992. *Maps of Meaning.* London: Routledge.

JAGER, B. 1985. "Body, House and City: The Intertwining of Embodiment, Inhabitation and Civilisation", in D. Seamon and R. Mugerauer (eds), *Dwelling, Place and Environment.* pp. 215-25. New York: Columbia University Press.

JAMBAI, A. and C. MacCORMACK 1996. Maternal Health, War, and Religious Tradition: Authoritative Knowledge in Pujehun District, Sierra Leone. *Medical Anthropology Quarterly* 10(2): 270-86.

JEDREJ, M. C. 1995. *Ingessana: The Religious Institutions of People of Sudan-Ethiopia Borderland.* Leiden: E. J. Brill.

JEDREJ, M. C. and R. SHAW (eds) 1992. *Dreaming, Religion and Society in Africa.* New York: E. J. Brill.

JESSEN, E. 1909. Nachrichten aus der Station Schira. *Evangelisch-Lutherisches Missionsblatt* 64(3): 60-2.

JINDRA, M. 2003. Natural/Supernatural Conceptions in Western Cultural Contexts. *Anthropological Forum* 13(2): 159-66.

JOHNS, T. and M. DUQUETTE 1991. Detoxification and Mineral Supplements as Functions of Geophagy. *American Journal of Clinical Nutrition* 53: 448-56.

JOHNSON, H. H. 1886. *The Kilima-njaro Expedition: A Record of Scientific Exploration in Eastern Equatorial Africa and a General Description of the Natural History, Languages and Commerse of the Kilima-njaro District.* London: Kegan Paul, Trench & Co.

JOHNSON, M. H. 1989. Conceptions of Agency in Archaeological Interpretation. *Journal of Anthropological Archaeology* 8: 189-211.

JOHNSON, P. H. 1946. Some Notes on Land Tenure on Kilimanjaro and the Vihamba of the Wachagga. *Tanganyika Notes and Records* 21: 1-20.

KALU, O. U. 1999. The Gods are to Blame: Religion, Worldview and Light Ecological Footprints in Africa. *Africana Marburgensia* 32(1-2): 3-27.

KELLY, R. L. 2003. "Colonization of New Land by Hunter-Gatherers: Expectations and Implications Based on Ethnographic Data," in M. Rockman and J. Steele (eds) *Colonization of Unfamiliar Landscapes: The Archaeology of Adaptation.* pp. 44-57. London: Routledge.

KERNER, D. O. 1995. "Chaptering the Narrative: The Material of Memory in Kilimanjaro, Tanzania," in M. C. Teski and J. J. Climo (eds) *The Labyrinth of Memory: Ethnographic Journeys.* pp. 113-27. Westport, CT: Greenwood Press.

KIERMAN, J. 1994. "Variations on a Christian Theme: The Healing Synthesis of Zulu Zionism," in C. Stewart and R. Shaw (eds) *Syncretism/Anti-Syncretism: The Politics of Religious Synthesis.* pp. 69-84. London: Routledge.

KIRK, G. S. 1970. *Myth: Its Meaning and Functions in Ancient and Other Cultures.* Cambridge: Cambridge University Press.

KIVUMBI, C. O. and W. D. NEWMARK 1991. "The History of the Half-Mile Strip on Mount Kilimanjro," in W. D. Newmark (ed.) *The Conservation of Mount Kilimanjaro.* pp. 81-6. Gland: IUCN.

KNUDSEN, J. W. 2002. Akula Udongo (Earth Eating Habit): A Social and Cultural Practice Among Chagga Women on the Slopes of Mount Kilimanjaro. *Indilinga: African Journal of Indigenous Knowledge Systems* 1: 19-26.

KOPONEN, J. 1988. *People and Production in Late Precolonial Tanzania: History and Structures.* Jyväskylä: Gummerus.

KORP, M. 2000. *Sacred Art of the Earth: Ancient and Contemporary Earthworks.* New York: Continuum.

KOPYTOFF, I. 1987. (ed.) *The African Frontier: The Reproduction of Traditional African Societies.* Bloomington: Indiana University Press.

KÜCHLER, S. 1993. "Landscape as Memory: The Mapping of Process and its Representation in a Melanesian Society," in B. Bender (ed.) *Landscape: Politics and Perspectives.* pp. 85-106. Oxford: Berg.

LAN, D. 1985. *Guns and Rain: Guerrillas and Spirit Mediums in Zimbabwe.* London: James Currey.

LAWUO, Z. E. 1984. *Education and Social Change in a Rural Community: A Study of Colonial Education and Local Response among the Chagga between 1920 and 1945.* Dar es Salaam: Dar es Salaam University Press.

LAYTON, R. and P. J. UCKO 1999. "Introduction: Gazing on the Landscape and Encountering the Environment," in P. J. Ucko and R. Layton (eds) *The Archaeology and Anthropology of Landscape: Shaping Your Landscape.* pp. 1-20. London: Routledge.

LEMA, A. A. 1968. The Lutheran Church's Contribution to Education in Kilimanjaro, 1893-1933. *Tanganyika Notes and Records* 68: 87-94.

LEMA, A. A. 1973. *The Impact of the Leipzig Lutheran Mission on the People of Kilimanjaro, 1893-1920.* Unpublished PhD thesis, University of Dar es Salaam.

LEMA, A. A. 2002. "Chaga Religion and Missionary Christianity on Kilimanjaro," in T. Spear and I. N. Kimambo (eds) *East African Expressions of Christianity.* pp. 39-62. Oxford: James Currey.

LEMAIRE, T. 1997. Archaeology between the Invention and Destruction of the Landscape. *Archaeological Dialogues* 4: 5-21.

LUBBREN, J. 2000. 'To Die Childless Means to Leave No Trace': Primary and Secondary Sterility among West African Women and its Individual and Social Implications. *Curare* 23(1): 41-9.

LUIG, U. 1995. "Naturaneignung als Symbolischer Prozeß in Afrikanischen Gesellschaften," in U. Luig and A. von Oppen (eds) *Naturaneignung in Afrika als Sozialer und Symbolischer Prozeß.* pp. 29-51. Berlin: Das Arabische Buch.

LUIG, U. and A. von Oppen (eds) 1995. *Naturaneignung in Afrika als Sozialer und Symbolischer Prozeß.* Berlin: Das Arabische Buch.

LUIG, U. and A. von OPPEN 1997. Landscape in Africa: Process and Vision. *Paideuma* 43: 8-45.

LUND, K. A. 1998. Landscape, Memory and Tourism in Southern Spain, Unpublished Ph.D. thesis, University of Manchester.

MacGAFFEY, W. 1990. Religion, Class, and Social Pluralism in Zaire. *Canadian Journal of African Studies* 24: 249-64.

MAGOWAN, F. 2001. Syncretism or Synchronicity? Remapping the Yolngu Feel of Place. *Australian Journal of Anthropology* 12(3): 275-90.

MAGOWAN, F. and J. GORDON 2001. Introduction. *Australian Journal of Anthropology* 12(3): 253-8.

MAKONI, S. 1998. "African Languages as European Scripts: The Shaping of Communal Memory," in S. Nuttall and C. Coetzee (eds) *Negotiating the Past: The Making of Memory in South Africa.* pp. 242-8. Oxford: Oxford University Press.

MALINOWSKI, B. 1974 (1925). *Magic, Science and Religion, and Other Essays.* London: Souvenir Press.

MAREALLE, P. I. 1947. *Maisha ya Mchagga Hapa Dunia na Ahera.* Nairobi: English Press.

MAREALLE, P. I. 1965. Chagga Customs, Beliefs and Traditions. *Tanganyika Notes and Records* 64: 56-61.

MAREALLE, T. L. M. 1952. The Wachagga of Kilimanjaro. *Tanganyika Notes and Records* 32: 57-64.

MARTIN, P. M. 1994. Contesting Clothes in Colonial Brazzaville. *Journal of African History* 35: 401-26.

MASAO, F. T. 1974. The Irrigation System in Uchagga: An Ethno-Historical Approach. *Tanzania Notes and Records* 75: 1-8.

MBITI, J. S. 1975. *Introduction to African Religion.* London: Heinemann.

McCALL, J. C. 1995. Rethinking Ancestors in Africa. *Africa* 65(2): 256-70.

McGLADE, J. 1999. "Archaeology and the Evolution of Cultural Landscapes: Towards an Interdisciplinary Research Agenda," in P. J. Ucko and R. Layton (eds) *The Archaeology and Anthropology of Landscape: Shaping Your Landscape.* pp. 458-82. London: Routledge.

MEYER, B. 1994. "Beyond Syncretism: Translation and Diabolization in the Appropriation of Protestantism in Africa", in C. Stewart and R. Shaw (eds) *Syncretism/Anti-Syncretism: The Politics of Religious Synthesis.* pp. 43-68. London: Routledge.

MEYER, H. 1891. *Across East African Glaciers: An Account of the First Ascent of Kilimanjaro.* (trans. E. H. S. Calder). London: Phillip & Son.

MNKENI, P. N. S. 1992. *Role of Soil Management in Enhancing Sustainability of Small-Holder Cropping Systems in some Agro-Ecosystems of Tanzania: A Review.* As: Agricultural University of Norway Press.

MOORE, S. F. 1970. Politics, Procedures, and Norms in Changing Chagga Law. *Africa: Journal of the International African Institute* 40(4): 321-44.

MOORE, S. F. 1976. The Secret of the Men: A Fiction of Chagga Initiation and its Relation to the Logic of Chagga Symbolism. *Africa* 46: 357-69.

MOORE, S. F. 1977. "Part I: The Chagga of Kilimanjaro." in W. M. O'Barr (ed) *The Chagga and Meru of Tanzania.* pp. 1-85. London: International African Institute.

MOORE, S. F. 1986. *Social Facts and Fabrications: 'Customary' Law on Kilimanjaro 1880-1980.* Cambridge: Cambridge University Press.

MOORE, S. F. 1996. Post-Socialist Micro-Politics: Kilimanjaro, 1993. *Africa* 66(4): 587-605.

MORPHY, H. 1993. "Colonialism, History and the Construction of Place: The Politics of Landscape in Northern Australia," in B. Bender (ed.), *Landscape: Politics and Perspectives.* pp. 205-43. Oxford: Berg.

MORPHY, H. 1995. "Landscape and the Reproduction of the Ancestral Past," in E. Hirsch and M. O'Hanlon (eds) *The Anthropology of Landscape: Perspectives on Place and Space.* pp. 184-209. Oxford: Oxford University Press.

MOSHA, R. S. 2000. *The Heartbeat of Indigenous Africa: A Study of the Chagga Educational System.* New York: Garland Publishing.

MOSKO, M. 2001. Syncretic Persons: Sociality, Agency and Personhood in Recent Charismatic Ritual Practices among North Mekeo (PNG). *Australian Journal of Anthropology* 12(3): 259-74.

MTURI, A. A. 1986. The Pastoral Neolithic of West Kilimanjaro. *Azania* 21: 53-63.

MUGERAUER, R. 2000. "Language and the Emergence of the Environment," in D. Seamon and R. Mugerauer (eds) 2000. *Dwelling, Place and Environment: Towards a Phenomenology of Person and World.* pp. 51-70. Malabar, FL: Krieger.

MULLIN, D. 2001. Remembering, Forgetting and the Invention of Tradition: Burial and Natural Places

103

in the English Early Bronze Age. *Antiquity* 75: 533-7.

MUNG'GONG'O, C. 1997. "Pangani Dam Versus the People," in A. D. Usher (ed.) *Dams as Aid: A Political Anatomy of Nordic Development Thinking.* pp. 105-18. London: Routledge.

MYERS, F. R. 1991. *Pintupi Country, Pintupi Self: Sentiment, Place, and Politics among Western Desert Aborigines.* Berkeley: University of California Press.

NEW, C. 1873. *Life, Wanderings and Labours in Eastern Africa: With Account of the First Successful Ascent of the Equatorial Snow Mountain, Kilima Njaro and Remarks upon East African Slavery.* London: Frank Cass & Co.

NTIRO, S. J. 1953. *Desturi za Wachagga.* Nairobi: Eagle Press.

NURSE, D. 1979. *Classifications of the Chaga Dialects: Language and History on Kilimanjaro, the Taita Hills and Pare Mountains.* Hamburg: Buske.

NURSE, D. and G. PHILIPPSON 1980. "The Bantu Languages of East Africa: A Lexicostatistical Survey," in E. C. Polome and C. P. Hill (eds) *Language in Tanzania.* pp. 26-67. London: Oxford International African Institute.

NUTTALL, S. and C. COETZEE (eds) 1998. *Negotiating the Past: The Making of Memory in South Africa.* Oxford: Oxford University Press.

O'BARR, W. M. 1977. "Introduction." in W. M. O'Barr (ed) *The Chagga and Meru of Tanzania.* pp. viii-xii. London: International African Institute.

ODNER, K. 1971. A Preliminary Report on an Archaeological Survey on the Slopes of Kilimanjaro. *Azania* 6: 131-49.

ODNER, K. 2004. "Bantu Ideology and Bantu Great Tradition," in T. Oestigaard, N. Anfinset and T. Saetersdal (eds) *Combining the Past and the Present: Archaeological Perspectives on Society.* pp. 119-26. Oxford: BAR Publishing.

OGUTU, M. A. 1972. The Cultivation of Coffee among the Chagga of Tanzania, 1919-1939. *Agricultural History* 46: 279-90.

O'KTING'ATI, A. and J. F. Kessy 1991. "The Farming Systems on Mt Kilimanjaro," in W. D. Newmark (ed.) *The Conservation of Mount Kilimanjaro.* pp. 71-80. Gland: IUCN.

OMARI, C. 1991. *God and Worship in Traditional Asu Society.* Erlangen: Verlag.

OTTO, R. 1950. *The Idea of the Holy.* Oxford: Oxford University Press.

PARRINDER, G. 1976. *African Traditional Religion.* Westport, Connecticut: Greenwood Press.

PEEL, J. D. Y. 1968. Syncretism and Religious Change. *Comparative Studies in Society and History* 10: 121-41.

PELS, P. 1999. *A Politics of Presence: Contacts between Missionaries and Waluguru in Late Colonial Tanganyika.* Amsterdam: Harwood Academic Publishers.

PETERSON, R. 2003. William Stukeley: An Eighteenth-Century Phenomenologist? *Antiquity* 77: 394-400.

PIKE, A. G. 1965. Kilimanjaro and the Furrow System. *Tanzania Notes and Records* 64: 95-6.

POLLARD, J. 2000. "Neolithic Occupation Practices and Social Ecologies from Rinyo to Clacton," in A. Richie (ed.) *Neolithic Orkney in its European Context.* Cambridge: McDonald Institute for Archaeological Research.

POLLARD, J. 2004. "A Movement of Becoming: Realms of Existence in the Early Neolithic of Southern Britain," in A. Chadwick (ed.) *Stories from the Landscape: Archaeologies of Inhabitation.* pp. 55-70. Oxford: BAR Publishing.

POLLARD, J. and M. Gillings 1998. Romancing the Stones: Towards a Virtual and Elemental Avebury. *Archaeological Dialogues* 5: 143-64.

POLT, R. 2000. *An Introduction to Heidegger.* London: Routledge.

POMEL, S. 1999. Soil Indicators of Anthropic Actions on Kilimanjaro. *IFRA Les Cahiers* 16: 74-8.

PRATT, M. L. 1992. *Imperial Eyes: Travel Writing and Transculturation.* London: Routledge.

RAINBIRD, P. 2002. "Marking the Body, Marking the Land: Body as History, Land as History: Tattooing and Engraving in Oceania," in Y. Hamilakis, M. Pluciennik and S. Tarlow (eds) *Thinking through the Body: Archaeologies of Corporeality.* pp. 233-47. London: Kluwer.

RANGER, T. O. 1997. Making Zimbabwean Landscapes: Painters, Projectors and Priests. *Paideuma* 43: 59-74.

RANGER, T. O. 1999. *Voices From the Rocks: Nature, Culture and History in the Matapos Hills of Zimbabwe.* Oxford: James Currey.

RAPOPORT, A. 1994. "Spatial Organisation and the Built Environment." in T. Ingold (ed.) *Cambridge Encyclopedia of Anthropology.* pp. 460-502. London: Routledge.

RASMUSSEN, S. J. 2002. Animal Sacrifice and the Problem of Translation: The Construction of Meaning in Tuareg Sacrifice. *Journal of Ritual Studies* 16(2): 141-64.

RAUM, O. F. 1940. *Chaga Childhood: A Description of Indigenous Education in an East African Tribe.* London: Oxford University Press.

REBMANN, J. 1860. "Journey to Jagga," in D. J. L. Krapf (ed.) *Travels, Researches, and Missionary Labours during an Eighteen Year's Residence in Eastern Africa. Together with Journeys to Jagga, Usambara, Ukambani, Shoa, Abessinia, and Khartum; and a Coasting Voyage from Mombaz to Cape Delgado.* pp. 230-65. London: Trubner and Co.

RENFREW, C. and C. SCARRE (eds) 1998. *Cognition and Material Culture: The Archaeology of Symbolic Storage.* Cambridge: McDonald Institute.

RICHARDS, C. 1991. "Skara Brae: Revisiting a Neolithic Village in Orkney," in W. A. Hanson and E. A. Slater (eds) *Scottish Archaeology: New Perceptions.* pp. 24-43. Aberdeen: Aberdeen University Press.

RICHARDS, C. 1996. Henges and Water: Towards an Elemental Understanding of Monumentality and Landscape in Late Neolithic Britain. *Journal of Material Culture* 1: 313-6.

RICH-DORMAN, S. 2002. 'Rocking the Boat'?: Churches, NGOs and Democratization in Zimbabwe. *African Affairs* 101: 75-92.

ROBBINS, J. 2002. My Wife Can't Break Off Part of Her Belief and Give it to Me: Apocalyptic Interrogation of Christian Individualism among the Urapmin of Papua New Guinea. *Padeuma* 48: 189-206.

ROBBINS, J. 2004. *Becoming Sinners: Christianity and Moral Torment in a Papua New Guinea Society.* Berkeley: University of California Press.

ROGERS, S. G. 1972. *The Search for Politcial Focus on Kilimanjaro.* Unpublished PhD thesis, University of Dar es Salaam.

ROSSINGNOL, J. and L. WANDSNIDER 1992. *Space, Time, and Archaeological Landscapes.* New York: Plenum Press.

SAHLINS, M. 1983. "Raw Women, Cooked Men and Other 'Great Things' of the Fiji Islands," in P. Brown and D. Tuzins (eds) *The Ethnography of Cannibalism.* pp. 72-93. Washington DC: Society for Psychological Anthropology.

SAHLINS, M. 1993. Goodbye to Tristes Tropes: Ethnography in the Context of Modern World History. *Journal of Modern History* 65: 1-25.

SANDGREN, D. P. 1989. *Christianity and the Kikuyu: Religious Divisions and Social Conflict.* New York: Peter Lang.

SCHAMA, S. 1996. *Landscape and Memory.* London: Fontana Press.

SCHUMAKER, L. 1997. "Constructing Racial Landscapes: Africans, Administrators, and Anthropologists in Late Colonia Northern Rhodesia," in P. Pels and O. Salemink (eds) *Colonial Subjects: Essays on the Practical History of Anthropology.* Ann Arbor: University of Michigan Press.

SCHUTZ, A. 1972. *The Phenomenology of the Social World.* London: Heinemann.

SERED, S. 2003. Afterword: Lexicons of the Supernatural. *Anthropological Forum* 13(2): 213-18.

SETEL, P. W. 1995a. *Youth, AIDS and the Changing Character of Adulthood in Kilimanjaro, Tanzania.* Unpublished PhD thesis, Boston University.

SETEL, P. W. 1995b. "The Social Context of AIDS Education among Young Men in Northern Kilimanjaro," in K-I Klepp, P. M. Biswalo and A. Talle (eds) *Young People at Risk.* pp. 49-68. Oslo: Scandinavian University Press.

SETEL, P. W. 1996. AIDS as a Paradox of Manhood and Development in Kilimanjaro, Tanzania. *Social Science and Medicine* 43(8): 1169-78.

SETEL, P. W. 1999. *A Plague of Paradoxes: AIDS, Culture and Demography in Northern Tanzania.* Chicago: University of Chicago Press.

SHANKS, M. and C. TILLEY 1987. *Social Theory and Archaeology.* Cambridge: Polity.

SHERIDAN, M. J. 2002. An Irrigation Intake Is like a Uterus: Culture and Agriculture in Precolonial North Pare, Tanzania. *American Anthropologist* 104(1): 79-92.

SHIPTON, P. 1994. Land and Culture in Tropical Africa: Soils, Symbols, and the Metaphysics of the Mundane. *Annual Review of Anthropology* 23: 347-77.

SIMMEL, G. 1990. *The Philosophy of Money.* London: Routledge.

SMEDJEBACKA, H. 1973. *Lutheran Church Autonomy in Northern Tanzania 1940-1963.* Åbo: Åbo Akademi.

SPEAR, T. 1996. "Struggles for the Land: The Political and Moral Economies of Land on Mount Meru," in G. Maddox, J. Giblin and I. N. Kimambo (eds) *Custodians of the Land: Environment and Hunger in Tanzanian History.* pp. 213-40. London: James Currey.

SPEAR, T. 1997. *Mountain Farmers: Moral Economies of Land and Agricultural Development in Arusha and Meru.* Oxford: James Currey.

SPEAR, T. and R. Waller (eds) 1993. *Being Maasai: Ethnicity and Identity in East Africa.* London: James Currey.

STAHL, G. E. 1708. *Theoria Medica Vera.* Halle: Literis Orpanotrophei.

STAHL, K. M. 1964. *History of the Chagga People of Kilimanjaro.* London: Mouton & Co.

STAHL, K. M. 1965. Outline of Chagga History. *Tanganyika Notes and Records* 64: 35-49.

STEWART, C. and R. SHAW (eds) 1994. *Syncretism/Anti-Syncretism: The Politics of Religious Synthesis.* London: Routledge.

STEWART, P. J. and A. STRATHERN 2003. "Introduction," in P. J. Stewart and A. Strathern (eds) *Landscape, Memory and History.* pp. 1-15. London: Pluto Press.

STEWART, R. and C. SHAW 1994. "Introduction: Problematizing Syncretism," in C. Stewart and R. Shaw (eds) *Syncretism/Anti-Syncretism: The Politics of Religious Synthesis.* pp. 1-26. London: Routledge.

STOCK, E. 1916. *The History of the Church Missionary Society.* London: Church Missionary Society.

STRATHERN, M. 1988. *The Gender of the Gift.* Berkeley, CA: University of California Press.

STRATHERN, M. 1991. *Partial Connections.* London: Roman and Littlefield.

STRATHERN, M. 1996. Cutting the Network. *Journal of the Royal Anthropological Institute* 2: 517-35.

SUNDKLER, B. G. M. 1961. *Bantu Prophets in South Africa.* London: Oxford University Press.

SWYNNERTON, R. J. M. 1949. Some Problems of the Chagga on Kilimanjaro. *East African Agricultural Journal* 15: 117-32.

TAÇON, P. S. C. 1994. Socialising Landscapes: The Long-Term Implications of Signs, Symbols and Marks on the Land. *Archaeology in Oceania* 29: 117-29.

TAGSETH, M. 2003. *Knowledge and Development in Mifongo Irrigation Systems.* Unpublished MPhil thesis, Norwegian University of Science and Technology, Trondheim.

TALLIS, R. 2004. *I Am: A Philosophical Inquiry into First-Person Being.* Edinburgh: Edinburgh University Press.

TARLOW, S. 2000. Emotions in Archaeology. *Current Anthropology* 41(5): 713-46.

THOMAS, J. 1996. *Time, Culture and Identity: An Interpretive Archaeology.* London: Routledge.

THOMAS, J. 1999. *Rethinking the Neolithic.* London: Routledge.

THOMAS, J. 2001. "Archaeologies of Place and Landscape," in I. Hodder (ed.) *Archaeological Theory Today.* pp. 165-86. Oxford: Polity.

THOMAS, J. 2004. *Archaeology and Modernity.* London: Routledge.

THORNTON, R. J. 1980. *Space, Time, and Culture among the Iraqw of Tanzania.* London: Academic Press.

THROOP, C. J. and K. M. MURPHY 2002. Bourdieu and Phenomenology. *Anthropological Theory* 2(2): 185-207.

TILLEY, C. 1994. *A Phenomenology of Landscape: Places, Paths and Monuments.* Oxford: Berg.

TILLEY, C. 1995. Rocks as Resources: Landscapes and Power. *Cornish Archaeology* 34: 5-57.

TILLEY, C. 1996. The Powers of Rocks: Topography and Monument Construction on Bodmin Moor. *World Archaeology* 28(2): 161-76.

TILLEY, C. 1999. *Metaphor and Material Culture.* Oxford: Blackwell.

TILLEY, C. 2000. "Materialism and an Archaeology of Dissonance," in J. Thomas (ed.) *Interpretive Archaeology: A Reader.* pp. 71-80. London: Leicester University Press.

TILLEY, C. 2004a. *The Materiality of Stone: Explorations in Landscape.* Oxford: Berg.

TILLEY, C. 2004b. Round Barrows and Dykes as Landscape Metaphors. *Cambridge Archaeological Journal* 14: 185-203.

TOKUNBO-WILLIAMS, P. 1997. Religion, Violence and Displacement in Nigeria. *Journal of Asian and African Studies* 32(1-2): 33-49.

TURNBULL, C. M. 1965. *Wayward Servants: The Two Worlds of the African Pygmies.* London: Eyre and Spottiswoode.

UCKO, P. J. 1969. "Ethnography and the Archaeological Interpretation of Funerary Remains" *World Archaeology* 1:262-80.

UCKO, P. J. and R. Layton (eds) 1999. *The Archaeology and Anthropology of Landscape: Shaping Your Landscape.* London: Routledge.

VAN DER VEER, P. 1994. "Syncretism, Multiculturalism and the Discourse of Tolerance," in C. Stewart and R. Shaw (eds) *Syncretism/Anti-Syncretism: The Politics of Religious Synthesis.* pp. 196-211. London: Routledge.

VAN DYKE, R. M. and S. E. Alcock 2003. "Archaeologies of Memory: An Introduction," in R. M. Van Dyke and S. E. Alcock (eds) *Archaeologies of Memory.* pp. 1-13. Oxford: Blackwell.

VAN GENNEP, A. 1965. *The Rites of Passage.* (1908). London: Kegan Paul.

VANSINA, J. 1966. *Kingdoms of the Savanna: A History of Central African States until European Occupation.* London: James Currey.

VANSINA, J. 1985. *Oral Tradition as History.* London: James Currey.

VANSINA, J. 1990. *Paths in the Rainforests: Toward a History of Political Tradition in Equatorial Africa.* London: James Currey.

VERMEER, D. E. and R. E. FERRELL 1985. Nigerian Geophagical Clay: A Traditional Antidiarrheal Pharmaceutical. *Science* 227: 634-6.

WAARDENBURG, J. (ed.) 1973. *Classical Approaches to the Study of Religion, Vol. 1.* Paris: Mouton.

WACH, J. 1951. *Types of Religious Experience, Christian and Non-Christian.* Chicago: University of Chicago Press.

WALSH, M. 1995. The Ritual Sacrifice of Pangolins among the Sangu of South-West Tanzania. *Bulletin of the International Committee on Urgent Anthropological and Ethnological Research* 37: 155-70.

WEINER, J. 1994. "Introduction", in J. Weiner (ed.) *Aesthetics in a Cross-Cultural* Category. Manchester: GDAT.

WEINER, J. 1996. Sherlock Holmes and Martin Heidegger. *Archaeological Dialogues* 3: 35-9.

WEISHAUPT, M. 1912. Überblick über unsere Missionsstationen in Ostafrika. *Evangelisch-Lutherisches Missionsblatt* 67(18): 430-3.

WEISS, B. 1993. Buying Her Grave: Money, Movement and AIDS in North-West Tanzania. *Africa* 63(1): 19-35.

WERBNER, R. 1989. *Ritual Passage, Sacred Journey.* Washington: Smithsonian Institution Press.

WERBNER, R. 1998. *Memory and the Postcolony: African Anthropology and the Critique of Power.* London: Zed Books.

WERBNER, R. and T. O. RANGER (eds) 1996. *Post-colonial Identities in Africa.* London: Zed Books.

WHEATLEY, D. 2004. Making Space for an Archaeology of Place. *Internet Archaeology* 15 http://intarch.ac.uk/journal/issue15/10/toc.html (last accessed 15th November 2005).

WHITE, L. 1993. Vampire Priests of Central Africa: African Debates about Labor and Religion in Colonial Northern Zambia. *Comparative Studies in Society in History* 35(4): 746-72.

WHITEHOUSE, H. 1995. *Inside the Cult: Religious Innovation and Transmission in Papua New Guinea.* Oxford: Oxford University Press.

WHITEHOUSE, H. 2000. *Arguments and Icons: Divergent Modes of Religiosity.* Oxford: Oxford University Press.

WHITEHOUSE, H. 2002. Modes of Religiosity: Towards a Cognitive Explanation of the Sociopolitical Dynamics of Religion. *Method & Theory in the Study of Religion* 14(7): 293-315.

WHITTLESEY, S. 1998. "Archaeological Landscapes: A Methodological and Theoretical Discussion," in S. Whittlesey, R. Ciolek-Torrello, and J. Altschul (eds) *Vanishing River.* Tucson, AZ: SRI Press.

WIESSNER, P. and A. TUMU 1998. *Historical Vines: Enga Networks of Exchange, Ritual, and Warfare in Papua New Guinea.* Washington, DC: Smithsonian Institution Press.

WILSON, M. and B. DAVID 2002. "Introduction," in B. David and M. Wilson (eds) *Inscribed Landscapes: Marking and Making Place.* pp. 1-9. Honolulu: University of Hawai'i Press.

WIMMELBÜCKER, L. 2003. *Kilimanjaro – A Regional History. Volume One: Production and Living Conditions, c. 1800-1920.* London: Lit Verlag.

WINTER, J. C. 1976. *Bruno Gutmann. 1876-1966: A German Approach to Social Anthropology.* Oxford: Clarendon Press.

WRIGHT, M. 1971. *German Missionaries in Tanganyika 1891-1941: Lutherans and Moravians in the Southern Highlands.* Oxford: Clarendon Press.

WYNN JONES, W. 1941. African Dugouts. *Tanganyika Notes and Records* 11: 11-2.

YNGSTROM, I. 2003. "Representations of Custom, Social Identity and Environmental Relations in Central Tanzania," in W. Beinart and J. McGregor (eds) *Social History and African Environments.* pp.175-95. Oxford: James Currey.

ZEDEÑO, M. N. and R. W. STOFFFLE 2003. "Tracking the Role of Pathways in the Evolution of a Human Landscape," in M. Rockman and J. Steele (eds) *Colonization of Unfamiliar Landscapes: The Archaeology of Adaptation.* pp. 59-80. London: Routledge.

ZELEZA, P. T. and E. KALIPENI (eds) 1999. *Sacred Spaces and Public Quarrels: African Cultural and Economic Landscapes.* Trenton: African World Press.

UNPUBLISHED SOURCES

TNA/KR/10/B/7/GEN/G. Tanzania National Archives, Kilimanjaro Region: 10/B/7/GEN/G. 24 January 1956. Letter from the Conservator of Forests, Northern Circle to the Provincial Forest Officer.

TNA/KR/15. Tanzania National Archives, Kilimanjaro Region: 15. 8 February 1958. Letter from the Assistant Director of Agriculture to the Conservator of Forests, Moshi.

TNA/KR/22/8/1. Tanzania National Archives, Kilimanjaro Region: 22/8/1. 3 January 1952. Letter from the Divisional Forest Officer, Northern Province to the District Commissioner Moshi.

TNA/KR/1953/8. Tanzania National Archives, Kilimanjaro Region: 1953/8. 20 November 1953. Handing over notes of the Rongai Forest Station.

TNA/NA/11/677. Tanzania National Archives, Native Affairs: 11/677. 4 July 1931. Letter from Provincial Commissioner Tabora to C. H. Gormley Acting District Commissioner.

TNA/NA/23/SF/1. Tanzania National Archives, Native Affairs: 23/SF/1. 24 April 1931. Letter from J. E. S. Lamb the District Officer to the Provincial Commissioner Tabora.

TNA/NA/20239/26. Tanzania National Archives, Native Affairs 20239/26: 30 January 1932. Letter from the Honorary Secretary of Native Affairs to an unspecified recipient.

TNA/NA/L5/21. Tanzania National Archives, Native Affairs: L5/21. 1957. Transcript of speech given by Paramount Chief Thomas Marealle on the opening of Moshi Art Exhibition.

TNA/NCNF/1930/185. Tanzania National Archives, Native Courts of the Nyanza Federation: Case 130. 28 October 1930. Case Ruling.

TNA/SE/AB988A. Tanzania National Archives, Secretariat: AB988A. 1922. Letter from the Senior Commissioner Moshi to the Chief Secretary to the Government, Dar es Salaam.

NOTES ON INTERVIEW TRANSCRIPTS

Copies of interview transcripts have been deposited in the University of Dar es Salaam Library, Dar es Salaam, Tanzania, the Library of the British Institute in Eastern Africa, Nairobi, Kenya, and the Library of the Department of Ethnography of the British Museum, London, England. Due to the sensitivity of many of the issues discussed and informants' wishes to protect their anonymity, and in accordance with the Royal Anthropological Institute's ethical guidelines, there is a 30 year embargo on their consultation from the date of fieldwork. The transcripts will thus be available for consultation from January 2035.

CAMBRIDGE MONOGRAPHS IN AFRICAN ARCHAEOLOGY

No 1   BAR S75, 1980   **The Niger Delta** *Aspects of its Prehistoric Economy and Culture* by Nwanna Nzewunwa. ISBN   0 86054 083 9

No 2   BAR S89, 1980   **Prehistoric Investigations in the Region of Jenne, Mali** *A Study in the Development of Urbanism in the Sahel* by Susan Keech McIntosh and Roderick J. McIntosh ISBN 0 86054 103 7

No 3   BAR S97, 1981   **Off-Site Archaeology and Human Adaptation in Eastern Africa** *An Analysis of Regional Artefact Density in the Amboseli, Southern Kenya* by Robert Foley. ISBN 0 86054 114 2

No 4   BAR S114, 1981   **Later Pleistocene Cultural Adaptations in Sudanese Nubia** by Yousif Mukhtar el Amin. ISBN 0 86054 134 7

No 5   BAR S119, 1981   **Settlement Patterns in the Iron Age of Zululand** *An Ecological Interpretation* by Martin Hall. ISBN 0 86054 143 6

No 6   BAR S139, 1982   **The Neolithic Period in the Sudan, c. 6000-2500 B.C.** by Abbas S. Mohammed-Ali. ISBN 0 86054 170 3

No 7   BAR S195, 1984   **History and Ethnoarchaeology in Eastern Nigeria** *A Study of Igbo-Igala relations with special reference to the Anambra Valley* by Philip Adigwe Oguagha and Alex Ikechukwu Okpoko. ISBN 0 86054 249 1

No 8   BAR S197, 1984   **Meroitic Settlement in the Central Sudan** *An Analysis of Sites in the Nile Valley and the Western Butana* by Khidir Abdelkarim Ahmed. ISBN 0 86054 252 1

No 9   BAR S201, 1984   **Economy and Technology in the Late Stone Age of Southern Natal** by Charles Cable. ISBN 0 86054 258 0

No 10   BAR S207, 1984   **Frontiers** *Southern African Archaeology Today* edited by M. Hall, G. Avery, D.M. Avery, M.L. Wilson and A.J.B. Humphreys. ISBN 0 86054 268 8. £23.00.

No 11   BAR S215, 1984   **Archaeology and History in Southern Nigeria** *The ancient linear earthworks of Benin and Ishan* by P.J. Darling. ISBN 0 86054 275 0

No 12   BAR S213, 1984   **The Later Stone Age of Southernmost Africa** by Janette Deacon. ISBN 0 86054 276 9

No 13   BAR S254, 1985   **Fisher-Hunters and Neolithic Pastoralists in East Turkana, Kenya** by John Webster Barthelme. ISBN 0 86054 325 0

No 14   BAR S285, 1986   **The Archaeology of Central Darfur (Sudan) in the 1st Millennium A.D.** by Ibrahim Musa Mohammed. ISBN 0 86054 367 6.

No 15   BAR S293, 1986   **Stable Carbon Isotopes and Prehistoric Diets in the South-Western Cape Province, South Africa** by Judith Sealy. ISBN 0 86054 376 5.

No 16   BAR S318, 1986 **L'art rupestre préhistorique des massifs centraux sahariens** by Alfred Muzzolini.. ISBN 0 86054 406 0

No 17   BAR S321, 1987   **Spheriods and Battered Stones in the African Early and Middle Stone Age** by Pamela R. Willoughby. ISBN 0 86054 410 9

No 18   BAR S338, 1987 **The Royal Crowns of Kush** *A study in Middle Nile Valley regalia and iconography in the 1st millennia B.C. and A.D.* by Lázló Török.. ISBN 0 86054 432 X

No 19   BAR S339, 1987   **The Later Stone Age of the Drakensberg Range and its Foothills** by H. Opperman. ISBN 0 86054 437 0

No 20   BAR S350, 1987 **Socio-Economic Differentiation in the Neolithic Sudan** by Randi Haaland. ISBN 0 86054 453 2

No 21   BAR S351, 1987   **Later Stone Age Settlement Patterns in the Sandveld of the South-Western Cape Province, South Africa** by Anthony Manhire. ISBN 0 86054 454 0

No 22   BAR S365, 1987 **L'art rupestre du Fezzan septentrional (Libye) Widyan Zreda et Tarut (Wadi esh-Shati)** by Jean-Loïc Le Quellec. ISBN 0 86054 473 7

No 23   BAR S368, 1987 **Archaeology and Environment in the Libyan Sahara** *The excavations in the Tadrart Acacus*, 1978-1983 edited by Barbara E. Barich. ISBN 0 86054 474 5

No 24   BAR S378, 1987   **The Early Farmers of Transkei, Southern Africa Before A.D. 1870** by J.M. Feely. ISBN 0 86054 486 9

No 25   BAR S380, 1987   **Later Stone Age Hunters and Gatherers of the Southern Transvaal** *Social and ecological interpretation* by Lyn Wadley. ISBN 0 86054 492 3

No 26   BAR S405, 1988 **Prehistoric Cultures and Environments in the Late Quaternary of Africa** edited by John Bower and David Lubell. ISBN 0 86054 520 2

No 27   BAR S418, 1988 **Zooarchaeology in the Middle Nile Valley** *A Study of four Neolithic Sites near Khartoum* by Ali Tigani El Mahi. ISBN 0 86054 539 3

No 28   BAR S422, 1988   **L'Ancienne Métallurgie du Fer à Madagascar** by Chantal Radimilahy. ISBN 0 86054 544 X

No 29   BAR S424, 1988 **El Geili The History of a Middle Nile Environment, 7000 B.C.-A.D. 1500** edited by I. Caneva. ISBN 0 86054 548 2

No 30   BAR S445, 1988   **The Ethnoarchaeology of the Zaghawa of Darfur (Sudan) Settlement and Transcience** by Natalie Tobert. ISBN 0 86054 574 1

No 31   BAR S455, 1988   **Shellfish in Prehistoric Diet Elands Bay, S.W. Cape Coast, South Africa** by W.F. Buchanan. ISBN 0 86054 584 9

No 32   BAR S456, 1988   **Houlouf I** *Archéologie des sociétés protohistoriques du Nord-Cameroun* by Augustin Holl. ISBN 0 86054 586 5

No 33   BAR S469, 1989 **The Predynastic Lithic Industries of Upper Egypt** by Liane L. Holmes. ISBN   0 86054 601 2 (two volumes)

No 34   BAR S521, 1989   **Fishing Sites of North and East Africa in the Late Pleistocene and Holocene** *Environmental Change and Human Adaptation* by Kathlyn Moore Stewart. ISBN 0 86054 662 4

No 35   BAR S523, 1989   **Plant Domestication in the Middle Nile Basin** *An Archaeoethnobotanical Case Study* by Anwar Abdel-Magid. ISBN 0 86054 664 0

No 36   BAR S537, 1989 **Archaeology and Settlement in Upper Nubia in the 1st Millennium A.D.** by David N. Edwards. ISBN 0 86054 682 9

No 37   BAR S541, 1989   **Prehistoric Settlement and Subsistence in the Kaduna Valley, Nigeria** by Kolawole David Aiyedun and Thurstan Shaw. ISBN 0 86054 684 5

No 38   BAR S640, 1996   **The Archaeology of the Meroitic State** *New perspectives on its social and political organisation* by David N. Edwards. ISBN 0 86054 825 2

No 39   BAR S647, 1996   **Islam, Archaeology and History** *Gao Region (Mali) ca. AD 900 - 1250* by Timothy Insoll. ISBN 0 86054 832 5

No 40   BAR S651, 1996 **State Formation in Egypt**: *Chronology and society* by Toby A.H. Wilkinson. ISBN 0 86054 838 4

No 41   BAR S680, 1997 **Recherches archéologiques sur la capitale de l'empire de Ghana** *Etude d'un secteur d'habitat à Koumbi Saleh, Mauritanie. Campagnes II-III-IV-V (1975-1976)-(1980-1981)* by S. Berthier. ISBN 0 86054 868 6

No 42   BAR S689, 1998 **The Lower Palaeolithic of the Maghreb** *Excavations and analyses at Ain Hanech, Algeria* by Mohamed Sahnouni. ISBN0 86954 875 9

No 43   BAR S715, 1998 **The Waterberg Plateau in the Northern Province, Republic of South Africa, in the Later Stone Age** by Maria M. Van der Ryst. ISBN 0 86054 893 7

No 44   BAR S734, 1998 **Cultural Succession and Continuity in S.E. Nigeria** *Excavations in Afikpo* by V. Emenike Chikwendu. ISBN 0 86054 921 6

No 45   BAR S763, 1999 **The Emergence of Food Production in Ethiopia** by Tertia Barnett. ISBN 0 86054 971 2

No 46   BAR S768, 1999 **Sociétés préhistoriques et Mégalithes dans le Nord-Ouest de la République Centrafricaine** by Étienne Zangato. ISBN 0 86054 980 1

No 47   BAR S775, 1999 **Ethnohistoric Archaeology of the Mukogodo in North-Central Kenya** *Hunter-gatherer subsistence and the transition to pastoralism in secondary settings* by Kennedy K. Mutundu. ISBN 0 86054 990 9

No 48   BAR S782, 1999 **Échanges et contacts le long du Nil et de la Mer Rouge dans l'époque protohistorique (IIIe et IIe millénaires avant J.-C.)** *Une synthèse préliminaire* by Andrea Manzo. ISBN 1 84171 002 4

No 49   BAR S838, 2000 **Ethno-Archaeology in Jenné, Mali** *Craft and status among smiths, potters and masons* by Adria LaViolette. ISBN 1 84171 043 1

No 50   BAR S860, 2000 **Hunter-Gatherers and Farmers** *An enduring Frontier in the Caledon Valley, South Africa* by Carolyn R. Thorp. ISBN 1 84171 061 X

No 51   BAR S906, 2000 **The Kintampo Complex** *The Late Holocene on the Gambaga Escarpment, Northern Ghana* by Joanna Casey. ISBN 1 84171 202 7

No 52   BAR S964, 2000 **The Middle and Later Stone Ages in the Mukogodo Hills of Central Kenya** *A Comparative Analysis of Lithic Artefacts from Shurmai (GnJm1) and Kakwa Lelash (GnJm2) Rockshelters* by G-Young Gang. ISBN 1 84171 251 5

No 53   BAR S1006, 2001 **Darfur (Sudan) In the Age of Stone Architecture c. 1000 - 1750 AD** *Problems in historical reconstruction* by Andrew James McGregor. ISBN 1 84171 285 X

No 54   BAR S1037, 2002 **Holocene Foragers, Fishers and Herders of Western Kenya** by Karega-Mûnene. ISBN 1 84171 1037

No 55   BAR S1090, 2002 **Archaeology and History in Ìlàrè District (Central Yorubaland, Nigeria) 1200-1900 A.D.** by Akinwumi O. Ogundiran. ISBN 1 84171 468 2

No 56   BAR S1133, 2003 **Ethnoarchaeology in the Zinder Region, Republic of Niger: the site of Kufan Kanawa** by Anne Haour. ISBN 1 84171 506 9

No 57   BAR S1187, 2003 **Le Capsien typique et le Capsien supérieur** *Évolution ou contemporanéité. Les données technologiques* by Noura Rahmani. ISBN 1 84171 553 0

No 58   BAR S1216, 2004 **Fortifications et urbanisation en Afrique orientale** by Stéphane Pradines. ISBN 1 84171 576 X

No 59   BAR S1247, 2004 **Archaeology and Geoarchaeology of the Mukogodo Hills and Ewaso Ng'iro Plains, Central Kenya** by Frederic Pearl. ISBN 1 84171 607 3

No 60   BAR S1289, 2004 **Islamic Archaeology in the Sudan** by Intisar Soghayroun Elzein. ISBN 1 84171 639 1.

No 61   BAR S1308, 2004 **An Ethnoarchaeological Study of Iron-Smelting Practices among the Pangwa and Fipa in Tanzania** by Randi Barndon. ISBN 1 84171 657 X.

No 62   BAR S1398, 2005 **Archaeology and History in North-Western Benin** by Lucas Pieter Petit. ISBN 1 84171 837 8.

No 63   BAR S1407, 2005 **Traditions céramiques, Identités et Peuplement en Sénégambie** *Ethnographie comparée et essai de reconstitution historique* by Moustapha Sall. ISBN 1 84171 850 5

No 64   BAR S1446, 2005 **Changing Settlement Patterns in the Aksum-Yeha Region of Ethiopia: 700 BC – AD 850** by Joseph W. Michels. ISBN 1 84171 882 3.

No 65   BAR S1454, 2006 **Safeguarding Africa's Archaeological Past** *Selected papers from a workshop held at the School of Oriental and African Studies, University of London, 2001* edited by Niall Finneran. ISBN 1841718920

No 66   BAR -S1537, 2006 **Excavations at Kasteelberg, and the Origins of the Khoekhoen in the Western Cape, South Africa** by Andrew B. Smith. ISBN 1 84171 969 2.

No 67   BAR –S1549, 2006 **Archéologie du Diamaré au Cameroun Septentrional** *Milieux et peuplements entre Mandara, Logone, Bénoué et Tchad pendant les deux derniers millénaires* by Alain Marliac ISBN 1 84171 978 1.

No 68   BAR –S1602, 2007 **Chasse et élevage dans la Corne de l'Afrique entre le Néolithique et les temps historiques** by Joséphine Lesur. ISBN 978 1 4073 0019 1.

No 69   BAR –S1617, 2007 **The Emergence of Social and Political Complexity in the Shashi-Limpopo Valley of Southern Africa, AD 900 to 1300** *Ethnicity, class, and polity* by John Anthony Calabrese ISBN 978 1 4073 0029 0.

No 70   BAR –S1658, 2007 **Archaeofaunal remains from the past 4000 years in Sahelian West Africa** *Domestic livestock, subsistence strategies and environmental changes* by Veerle Linseele ISBN 978 1 4073 0094 8.

No 71   BAR –S1667, 2007 **Il Sahara centro-orientale Dalla Preistoria ai tempi dei nomadi Tubu / The Central-Oriental Sahara. From Prehistory to the times of the nomadic Tubus** by Vanni Beltrami con le fotografie e i riassunti in inglese di Harry Proto / with English summaries and photographs by Harry Proto. ISBN 978 1 4073 0102 0.

www.ingramcontent.com/pod-product-compliance
Lightning Source LLC
Chambersburg PA
CBHW061003030426
42334CB00033B/3345

* 9 7 8 1 4 0 7 3 0 1 1 7 4 *